Proceedings of the 30th International Geological Congress
Volume 6

Global Tectonic Zones
Supercontinent Formation and Disposal

Proceedings of the 30th International Geological Congress

Proceedings of the
30th International
Geological Congress

Beijing, China, 4 - 14 August 1996

VOLUME 6

Global Tectonic Zones Supercontinent Formation and Disposal

Editors:

Xiao Xuchang
Institute of Geology, CAGS, Beijing, China
Liu Hefu
China University of Geosciences, Beijing, China

CRC Press
Taylor & Francis Group
Boca Raton London New York

CRC Press is an imprint of the
Taylor & Francis Group, an **informa** business

First published 1997 by VSP BV

Published 2019 by CRC Press
Taylor & Francis Group
6000 Broken Sound Parkway NW, Suite 300
Boca Raton, FL 33487-2742

First issued in paperback 2019

No claim to original U.S. Government works

ISBN 13: 978-0-367-44812-7 (pbk)
ISBN 13: 978-90-6764-262-0 (hbk)

This book contains information obtained from authentic and highly regarded sources. While all reasonable efforts have been made to publish reliable data and information, neither the author[s] nor the publisher can accept any legal responsibility or liability for any errors or omissions that may be made. The publishers wish to make clear that any views or opinions expressed in this book by individual editors, authors or contributors are personal to them and do not necessarily reflect the views/opinions of the publishers. The information or guidance contained in this book is intended for use by medical, scientific or health-care professionals and is provided strictly as a supplement to the medical or other professional's own judgement, their knowledge of the patient's medical history, relevant manufacturer's instructions and the appropriate best practice guidelines. Because of the rapid advances in medical science, any information or advice on dosages, procedures or diagnoses should be independently verified. The reader is strongly urged to consult the relevant national drug formulary and the drug companies' and device or material manufacturers' printed instructions, and their websites, before administering or utilizing any of the drugs, devices or materials mentioned in this book. This book does not indicate whether a particular treatment is appropriate or suitable for a particular individual. Ultimately it is the sole responsibility of the medical professional to make his or her own professional judgements, so as to advise and treat patients appropriately. The authors and publishers have also attempted to trace the copyright holders of all material reproduced in this publication and apologize to copyright holders if permission to publish in this form has not been obtained. If any copyright material has not been acknowledged please write and let us know so we may rectify in any future reprint.

Visit the Taylor & Francis Web site at
http://www.taylorandfrancis.com

and the CRC Press Web site at
http://www.crcpress.com

CONTENTS

Proc. 30ᵗʰ Int'l. Geol. Congr., Vol. 6, pp. 1-14
Xiao Xuchang and Liu Hefu (Eds)
© VSP 1997

Tectonic Framework and Geodynamic Evolution of Eastern China

REN JISHUN, NIU BAOGUI , HE ZHENGJUN, XIE GUANGLIAN AND LIU ZHIGANG
Institute of Geology, Chinese Academy of Geological Sciences, Beijing 100037, China

Abstract

The tectonics of eastern China in the Paleozoic was mainly under the control of the Paleo-Asian Ocean dynamic system and the tectonic belts run nearly E-W. During the Triassic-Jurassic period, it was under the joint control of the Tethys and Paleo-Pacific systems, some tectonic belts turning to NE, and the others keeping to nearly E-W. By the time of the Late Jurassic-initial Cretaceous the collision between West Pacifica (the West Pacific continent) and Asia resulted in the formation of the marginal orogen of eastern Asia and the epicontinental activated belt of eastern China and the E-W tectonic framework was completely replaced by the NE-NNE framework. So the turn of the tectonic regime and dynamic regime in eastern China began in the Indosinian orogeny (Late Triassic) and was finally completed in the Yanshanian orogeny (Late Jurassic-initial Cretaceous). From the Cretaceous on, the tectonics of eastern China was mainly under the control of the Pacific system and, as West Pacifica disintegrated and submerged and the Pacific was formed, eastern China rifted on a large scale, leading to the formation of the rift basin system of eastern Asia and the trench-arc-basin system of the western Pacific.

Keywords: eastern China, framework, Paleo-Pacific, West Pacifica, epicontinental activated belt, Tethys

INTRODUCTION

There are two tectonic systems which are different from each other in age, nature and character, one E-W striking, constructing the basis of the tectonics of eastern China and basically reflecting the tectonic feature of the Paleozoic and, the other NE-NNE striking, superimposed on the former and basically showing the tectonic feature of the Mesozoic and Cenozoic. These two tectonic systems, becoming intersected and superimposed, give an extremely complicated tectonic picture in the geological and tectonic maps of China.

How was the E-W striking tectonic framework transformed into the NE-NNE striking one? What did the process look like? How still was the dynamic system transformed? In the 1970's, we pointed out that the great importance of the Indosinian movement is in that it is a great turning point in the development of the crustal structure, it broke the Paleozoic tectonic framework of eastern China and commenced the development of the Marginal Pacific tectonic domain [17, 21]. After further studies we pointed out again in the 1980's that the Mesozoic orogeny which began in the Indosinian and reached its climax in the Yanshanian is considered to be the most important tectonic transformation in the Phanerozoic in eastern China [32, 33]. We re-studied the tectonics of eastern China in the 1990's and we not only confirmed our original point of view but also obtained some new ideas. Here is our report. We look forward to your critical remarks.

NEARLY E-W STRIKING TECTONIC UNITS (Fig. 1)

From north to south, the nearly E-W striking tectonic units are:
Eastern section of the Tianshan-Hing'an Orogenic System
Sino-Korean Paraplatform

2

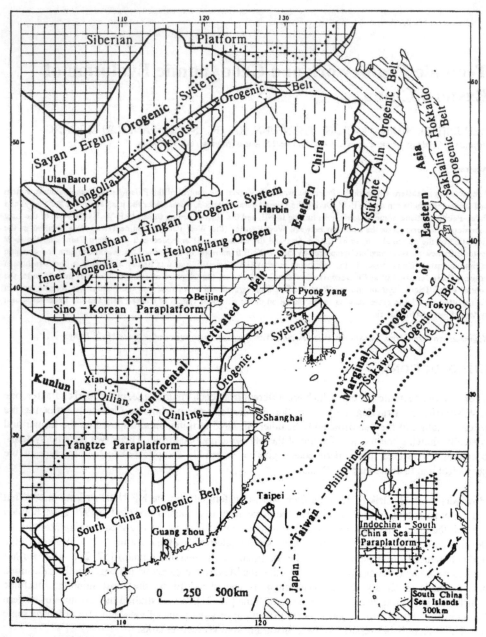

Figure1. **Tectonic divisions of eastern China and adjacent areas**

Eastern section of the Kunlun-Qilian-Qinling Orogenic System
Yangtze Paraplatform
South China Orogenic Belt
Indochina-South China Sea Paraplatform

There are the Sayan-Ergun orogenic system, the Siberian platform, etc. north of the Tianshan-Hing'an orogenic system, but we will not deal with them in detail because the most parts of them are outside of China.

Eastern Section of the Tianshan-Hingan Orogenic System
It includes the South Mongolia-Hing'an Variscan orogenic belt, the Ondor Sum Caledonian orogenic belt, the Jilin-Heilongjiang block (containing the Songliao, Burya-Jiamusi, Xingkai blocks), the Inner Mongolia-Jilin-Heilongjiang Indosinian orogenic belt, etc.

It was generally held in the past that the orogenesis in the Tianshan-Hing'an orogenic system concluded in the Variscan movement and thus it was called the Variscides [18, 22, 29, 34]. It seems now that the Tianshan-Hing'an orogenic system of the Paleo-Asian Ocean was certainly completed by the Maokou stage of the Permian as is marked by the unconformity under the Zhesi Formation of the Maokou stage which also means the start of a new stage.

Based mainly on the distinguishing, by the Regional Geological Survey Party of Jilin, of the Early Triassic Lujiatun Formation from the former folded Permian sequence [6], the postulation that the Linxi and Taohaiyingzi formations of Inner Mongolia may include Triassic strata and the fact that the folded Permian and Triassic are continuous in Far East Russia [23], we put forward in 1984 that the former Inner Mongolia-Jilin-Heilongjiang Variscan foldbelt should be an Indosinian orogenic belt, the main orogeny occurring by the Late Triassic[37]. Now this proposal has been fully verified. After detailed study we have found Triassic Ostracoda and Conchostraca in the Linxi Formation in Inner Mongolia[1], confirming that it does contain Triassic horizons. We have also discovered that the lower part of the Permian-Triassic Linxi Formation is mainly composed of marine sediments and includes turbidite [14]. The Geological Map of Heilongjiang Province shows a vast area covered by Permian-Triassic or Permian-Lower Triassic sequence, suggesting that the Permian in the Inner Mongolia-Jilin-Heilongjiang region is continuous to the Triassic, and therefore, the tectonic maps of both Jilin and Heilongjiang provinces mark the area as an Indosinian orogenic belt [1, 2]. All these prove that there did exist a generally E-W or NNE striking Indosinian orogenic belt in the Inner Mongolia-Jilin-Heilongjiang region, which was superimposed on the Paleozoic tectonic units in this area and extended westward to the Beishan region in Gansu and eastward to Sikhote Alin in Russia to become the basement of the Sikhote Alin Yanshanian orogenic belt.

The Inner Mongolia-Jilin-Heilongjiang orogenic belt suffered again in the intense Yanshanian orogeny in the Middle-Late Jurassic, in which the crust was further shortened. This is evidenced by: 1) in the Beishan Mts. of Gansu, the border region between China and Mongolia, Inner Mongolia, Jilin, Heilongjiang, etc., there are large-scale thrusts cutting into the Triassic and Early-Middle Jurassic and covered by the Late Jurassic-Early Cretaceous Greater Hing'an Volcanics and the Cretaceous Bayanhua Formation [1, 3, 4, 43]; 2) in the Mishan-Jixi region of Heilongjiang, the mylonite in the nearly E-W striking thrusts yielded Ar^{39}/Ar^{40} plateau age of 167-169Ma (Li Jinyi and Niu Baogui, in press); and 3) the mylonite in the metamorphic core complex that marked the extension after the orogeny in the border region between China and Mongolia gave an Ar^{39}/Ar^{40} plateau age of 155.1 ± 10Ma [43]. Thus the orogeny in the Inner Mongolia-Jilin-Heilongjiang

---

1) Ostracoda fossils were determined by Li Yougui and the Conchostraca by Liu Shuwen.

orogenic belt came to an end in the Middle-Late Jurassic or 170-155Ma.

It is to be especially pointed out that the Mesozoic thrust system and magmatic belt extending from the Yinshan Range in Inner Mongolia to the Yanshan region in Hebei seem to be a part of the Inner Mongolia-Jilin-Heilongjiang orogenic belt and form its southern margin.

Sino-Korean Paraplatform

It is the oldest and most well-known Paleoproterozoic platform within the territory of China. Because of its small size it is more active than other cratons and thus it is called a paraplatform [19]. The Sino-Korean paraplatform is the main part of the Sino-Korean plate. The oldest rock in its basement has radiometric age of 3800Ma [28]. In the formation of the platform the tectono-thermal events at 3000, 2800, 2600, 2400 and 1900-1700Ma are especially important [40]. The widespread appearing of the potassium granite in 2350Ma marked the final formation of the crystallized basement of the paraplatform and its general cratonization. Then, after the Zhongtiao (or Luliang) orogenic cycle in 1900-1700Ma, the platform eventually cratonized.

Eastern Section of the Kunlun-Qilian-Qinling Orogenic System

Up to now, the academic community has generally had a consensus: the eastern section of the Kunlun-Qilian-Qinling orogenic system extends to the Dabie Mts. and still eastward appears in the North Jiangsu-Jiaonan area across the Tangcheng-Lujiang transform fault, forming an important orogenic belt in eastern Asia, which runs across the Yellow Sea to appear in the Korean Peninsula.

Some geologists hold that the Qinling oceanic basin between the Sino-Korean and Yangtze paraplatforms was existing from the Proterozoic to the Triassic. Therefore they regard the Qinling orogen as an Indosinian collisional one [16, 39, 42]. But we firmly believe that the Qinling orogenic belt had a polycyclic development by stages. During the Meso-Neoproterozoic, the Qinling proto-ocean separated the Sino-Korean and Yangtze blocks and disappeared at 800Ma followed by a continent-continent collisional orogeny, resulting in the unification of the Sino-Korean, Yangtze and other blocks to form the Chinese protoplatform [34]. After the Early Cambrian, the Chinese protoplatform disintegrated and small oceanic basin was formed in the Qilian region and probably also in the Qinling area. But after the Caledonian orogeny, pelagic sediments had never appeared again in the Qinling area. The ultra-mafic rocks believed to be ophiolite are all of pre-Devonian age. There was the small ocean basin of the Late Permian-Early Triassic Anyemaqen Tethys, but it did not extend to the eastern Qinling area. The recent discovery of radiolarians probably of Triassic age in the Tongbai area [11] suggests that the water of the Anyemaqen basin did run eastward to the Tongbai-Dabie region, but not necessarily that there then existed a deep oceanic basin because the rocks containing the fossils are not radiolarian cherts co-existing with ophiolite but radiolarian-bearing carbonates (limestones). Thus we say the Qinling is not a simple Indosinian collisional orogenic belt but a composite polycyclic one. The Variscan, Indosinian and Yanshanian orogenies are all not continent-continent collisional orogeny following the disappearing of the oceanic basin but continent-continent imbrication orogeny and strike-slipping orogeny long after the disappearing of the ocean. The enormously thick Middle Jurassic red molasse in the piedmont of the Daba Mts. [13] shows that there still was important orogeny in the Jurassic in the Qinling region and the formation of the Cretaceous red basins in the eastern and western Qinling [41] suggests a strike-slipping extension regime after the orogeny. All these demonstrate that the Qinling orogenic belt was not completed until the Middle and Late Jurassic or even initial Cretaceous.

Yangtze Paraplatform

It was finally formed at 800Ma. Lying on its basement are the cover of the Sinian-Triassic marine sediments. Since the Late Triassic sediments on it are mainly terrestrial. The Yangtze paraplatform constitutes the main part of the Yangtze plate (or South China plate). Its basement is obviously two-layered, the lower layer being a crystallized basement formed in the Archean-Paleoproterozoic (1800Ma) and the upper being a folded basement formed in 1000-800Ma. Under the basins of Sichuan, Jianghan, and North Jiangsu-South Yellow Sea there is no folded basement or it is very thin and the cover usually directly rests on the crystallized basement. In the rest of the platform the crystallized basement is buried deep under the folded basement.

South China Orogenic Belt

After more than a decade of repeated study and debate, it has now firmly been demonstrated that the "Banxi Group" is not a melange and South China is not a Mesozoic collisional orogenic belt. In the Sinian-Early Paleozoic it was an aulacogen based on the continental crust, a branch of the Tibet-Malay-South China triple juncture [35]. The Caledonian continent-continent imbrication orogeny made it folded and converted it into a post-Caledonian platform on which developed the Devonian-Triassic sedimentary cover. In the Mesozoic, because of the activity of the Tethys-Paleo-Pacific plate there occurred intense tectono-magmatism resulting in a very complicated fold-fault system development, widespread granite intrusion and large-scale volcanism, as well as in occurrence of plentiful endogenetic metallic deposits, especially the well-known tungsten-tin deposits and rare earth deposits of niobium, tantalum, etc.

Indochina-South China Sea Paraplatform

The conception that there existed a continental block in coastal southeast China was first put forward by Grabau [12], who called it Cathaysia. After careful analysis on the coastal metamorphic sequence, according to which Grabau conceived Cathaysia, Huang pointed out that Cathaysia should be a Caledonian orogenic belt or a post-Caledonian massif [18]. In 1964, Ren Jishun, based on his regional investigation and paleogeographical analysis in South China, proposed that there was a pre-Sinian landmass south of the South China Caledonian foldbelt, and that it was this landmass that provided the debris to form the enormously thick Sinian-Silurian sequence in South China, and it was called Indochina-South China Sea platform [34, 36]. After decades of re-investigations and drillings and geophysical studies on the seas[1], especially the paleogeographical analysis of Liu Baojun et al., it has fully been demonstrated that under the present seas there did exist a large landmass in the past [27]. The coastal metamorphic series, such as the Chencai Group in eastern Zhejiang, the Jian'ou Group in northern Fujian and the Yunkai Group in western Guangdong, are the rocks in the root of the South China orogenic belt that were deeply involved in the Caledonian orogeny (Jin Wenshan et al., in press).

NE-NNE STRIKING STRUCTURAL BELTS

There are two successive systems, the marginal orogenic system of eastern Asia and the rifted basin system (or the basin-range system) of eastern Asia, the latter including the trench-arc-basin system of the western Pacific.

Marginal Orogenic System of Eastern Asia

It consists of the marginal orogenic belt of eastern Asia and the epicontinental activated belt of eastern China.

Marginal Orogenic Belt of Eastern China

A part of the global circum-Pacific orogenic belt, the marginal orogenic belt of eastern Asia runs

---

1) unpublished data of the Shanghai Bureau of Marine Geology and Guangzhou Bureau of Marine Geology

from northeastern Asia through Japan to the Indonesian archipelago, connecting with the Cordilleran orogenic belt of North America in the north and with the New Guinea orogenic belt north of Australia in the south. The tectonic units of the marginal orogenic belt of eastern Asia that are more closely related to the tectonic evolution of eastern China are the Mongolia-Okhotsk orogenic belt, the Sikhote Alin orogenic belt, the Sakawa orogenic belt of Japan and the Japan-Taiwan-Philippines arc.

Mongolia-Okhotsk Orogenic Belt It is an orogenic belt developed from the Paleozoic all the way to the Mesozoic. The Paleozoic is mainly a volcani-silico-terrigenous formation, and probably ophiolite, and the Mesozoic (Triassic-Jurassic) is mainly a terrigenous clastic flysch. Great breaks of tectono-magmatic cycle are present in the Upper Devonian-Lower Carboniferous, Upper Triassic and Jurassic-Cretaceous, suggesting a polycyclic development [44]. In its evolution, the marine basin closed gradually from west to east and the orogeny migrated from Mongolia to Okhotsk and came to an end in end-Jurassic-initial Cretaceous. It is the belt where Siberia and the Chinese blocks finally welded. This development had a great influence on the northeastern part of China.

Sikhote Alin Orogenic Belt Connecting to the Koriaka orogenic belt to the north and next to the Sakhalin-Hokkaido orogenic belt to the east, it is a Late Yanshanian orogenic belt based on Indosinian orogenic belt. It is mainly composed of a Late Triassic-Late Cretaceous volcani-silico-sandy-argilaceous formation with some volcanic and basic-ultrabasic rocks, but , it is a problem of future study whether there is real ophiolite representing oceanic crust. There occurred three major orogenic movements, the first in the end-Jurassic-initial Cretaceous, marked by the unconformity under the Berriasian-Valanginian Formation, the second in the Early Cretaceous Hauterivian marked by the unconformity under the Baliemu Formation, and the third after the Sartanian and before the Mastrichtian. These three events are roughly equivalent to those represented by the unconformities under the Zhangjiakou Formation and the Yixian Formation, the Denglouku Formation and the Sifangtai Formation respectively. Some workers took this orogenic belt as a Mesozoic accretional terrane [31, 38], but, it seems that the evidences are not sufficient, because there are neither authentic remains of the Mesozoic oceanic crust nor record of consumed oceanic crust material. It is extremely unsafe to determine the tectonic nature based only on the analysis of fossils.

Sakawa Orogenic Belt of Japan It is a pre-Cretaceous orogenic belt of Japan. It is called the Sakawa orogenic belt because its main orogeny is the well-known Sakawa movement. As it is the best-documented and best-studied region in the marginal orogenic belt of eastern Asia it can be taken as the typical of this belt.

According to Ichikawa et al. [20], we can make some judgment on the evolution of the Sakawa orogenic belt:

The Paleo-Pacific in Japan subducted under the Asian continent in the Middle-Late Triassic, starting the Akiyoshi (Indosinian) orogeny which formed the Sangun metamorphic belt and the Akiyoshi orogenic belt (including the Akiyoshi and Maizuru terranes) southeast of the Hida block. The Sakawa orogeny in the Late Jurassic-initial Early Cretaceous is the most important orogenic cycle in this belt in that it marked the closing of the Paleo-Pacific and the collision of Asia with West Pacifica, resulting in the great Yanshanian orogenic belt along the eastern margin of Asia and the magnificent epicontinental activated belt of eastern China.

The Sakawa orogenic belt extends to northeastern Asia through Sikhote Alin and Hokkaido-Sakhalin islands and southward to the Philippines through Ryukyu-Taiwan islands. There are differences in these three regions. In southwest Japan, the younger orogenic belts migrated outward after the Sakawa orogeny, leading to the formation of the Cenozoic Shimanto orogenic belt. But, in Sikhote Alin and Sakhalin-Hokkaido, no mountain was formed after the orogeny, instead, there appeared a new marine basin under the new extending regime and the overall Mesozoic orogeny did not complete until the end of Cretaceous. In the Ryukyu-Taiwan region, the great Mesozoic orogenic belt formed in the Sakawa orogeny was profoundly remolded because of the occurrence of the Tertiary-Quaternary island arc-marginal sea (Okinawa trough), but its existence can be known from the intense tectono-magmatism in the Jurassic-initial Cretaceous in the hanging wall of the plate collision belt, the Korean Peninsula-coastal southeastern China.

What is to be pointed out is that many a few workers took the Mesozoic orogenic belt of Japan as an accretional orogenic belt formed by the subduction of the Kula plate under the Asian continent. In actual fact, the so-called Kula plate was conceived based on only a few paleomagnetic data [15] and is not supported by more practical materials. In contrast, more and more facts make us believe in the idea that the Philippines, Okhotsk and even the western Pacific were once land [8, 9, 30, 33]. So we regard the Mesozoic marginal orogenic belt of eastern Asia represented by the Sakawa orogen of Japan as a collisional orogenic belt, i.e., the collisional orogenic belt between West Pacifica and the Asian continent. Actually some French geologists, such as J. Charvet et al., have offered eloquent evidences. Based on their field investigations in Japan, they clearly pointed out that the Sakawa orogenic belt is the result of the collision between the landmass of south Japan and the Asian continent [5, 10]. We support these French scientists and hold that the landmass of south Japan is actually a part of the huge landmass of West Pacifica.

Epicontinental Activated Belt of Eastern China
The term epicontinental activated belt was coined in 1980 [34]. It is to stressed that an epicontinental activated belt is different from an active continental margin. The active margin is a zone under direct action of plate margin and belongs to an orogenic belt or a geosynclinal foldbelt, whereas the epicontinental activated belt is a zone of more or less intense tectono-magmatism resulting from the reactivation of the stable continental crust landward of the active margin under the action of plate margin. The Mesozoic epicontinental activated belt of eastern China is a typical example.

The epicontinental activated belt of eastern China is a vast area east of Lake Baikal-Ordos basin-Sichuan basin and west of the Sakawa orogenic belt of Japan. In some parts of area orogenic processes had been completed and continental crust stabilized in the Paleozoic and in the others cratonization fulfilled before the Sinian, but in the Mesozoic they all, under one and the same tectonic action, were reactivated and formed a new tectonic belt superimposed on the older tectonic units. The epicontinental activated belt of eastern China and the marginal orogenic belt of eastern Asia are two important units that came into being at the same time and co-existed. It is both of them that constitute the great Mesozoic mountain ranges of eastern Asia similar to the present-day Cordillera orogenic belt of North America.

In the past, geologists usually related the Mesozoic tectono-magmatism of eastern China to the activity of the Pacific [24, 25, 34]. It seems now that, however, it is in fact the result of the joint action, on eastern China, of the Paleo-Pacific of Japan (the main branch of the Paleo-Pacific between of West Pacifica and the Asian continent) and the Mongolia-Okhotsk Paleo-Pacific (an aulacogen-type rift of the Paleo-Pacific cutting into the Asian continent, which was located between Siberia and eastern China) in their evolution.

The epicontinental activated belt of eastern China between the Mongolia-Okhotsk belt and the Sakawa orogenic belt of Japan can be divided into three en enchlon sub-belts from west to east: the Inner Mongolia-Hing'an belt or the western belt including Inner Mongolia, Greater and Lesser Hing'an Ranges, etc.; the central Sino-Korean and Yangtze platform or the central belt encompassing the middle section of the Yangtze paraplatform between the Xuefengshan Mts. and the Sichuan basin; and the central portion of the Sino-Korean paraplatform between the Tancheng-Lujiang fault and the Ordos basin and the South China-Lower Yangtze-Jiaoliao belt or the eastern belt including the region east of the Xuefengshan Mts. and the Tancheng-Lujiang fault and the Korean Peninsula.

In the eastern belt the tectono-magmatic reactivation started in the Late Triassic Indosinian orogeny and reached its acme in the Jurassic Yanshanian orogeny, resulting in the NE-NNE tectono-magmatic belt. The NE striking folding and faulting system cut across the old E-W boundary between the Yangtze paraplatform and South China orogenic belt. It is the most conspicuous in the Jiulingshan Mts. and further east. Sedimentologically the Sinian-Triassic is generally the same in the Lower, Middle and Upper Yangtze and is all the cover of the Yangtze paraplatform, but there is difference in the Mesozoic tectono-magmatism in the three regions. In the Jiulingshan and the Lower Yangtze the tectono-magmatism started in the Late Triassic Indosinian cycle, the cover folding taking place before the sedimentation of the Jurassic Xiangshan Group and in the Jurassic-initial Cretaceous Yanshanian orogeny there occurred strong detachment in the deep of the crust and thrusting and folding in the upper layer, and large-scale strike-slipping volcanic activity and granite intrusion, whereas in the Middle and Upper Yangtze the cover folding did not take place until the Middle Cretaceous (ca. 100Ma).

In the central belt the tectono-magmatic reactivation seems to be much weaker than in the eastern belt, which was mainly marked by the detachment in the deep of the crust, cover folding-thrusting and local magmatism. Generally there was no magmatism in the middle portion of the Yangtze platform. In the middle portion of Sino-Korea, magmatism was concentrated in the northern Taihangshan Mts. and the Yanshan-Liaoning region near Inner Mongolia and tectonic transformation occurred relatively late. In the Hunan-Guizhou area east of the Sichuan basin the cover folding took place before the sedimentation of the Middle-Late Cretaceous Qianjiang Formation. In the Shanxi and Hebei region east of the Ordos basin the cover folding and faulting took place after the Middle Jurassic but before the Early Cretaceous

In the western belt the tectono-magmatic reactivation in Inner Mongolia, the Greater and Lesser Hing'an Ranges started also in the Indosinian cycle and reached its climax in the Yanshanian stage characterized by the large scale crustal detachment in the deep, thrusting, strike-slipping and strong K-rich calc-alkaline magmatism. It is especially to be pointed out that the tectono-magmatic reactivation in this region was not related to the activity of the Paleo-Pacific of Japan but to the formation and development of the Mongolia-Okhotsk belt.

The Rifting Basin System of Eastern Asia (Basin-Range System of Eastern Asia)
The formation of the complicated fold-fault system and K-rich calc-alkaline volcanics and enormous granite resulting from the Yanshanian orogeny in the Jurassic-initial Cretaceous marked the conclusion of the evolution of a tectonic regime, the Paleo-Pacific. Then since the Cretaceous (starting in 140Ma), especially the Middle Cretaceous (120 or 100Ma), eastern Asia has been in an overall environment of tension-shearing which produced a unique basin-range system, the rift basin system of eastern Asia and the trench-arc-basin system of the West Pacific, marking a new stage, the formation and development of a new oceanic basin, the Pacific ocean.

The rift basin system in different parts of eastern China and eastern Asia has different styles and characteristics depending on the one hand on the dynamic condition of its formation and on the other hand on the older tectonic background under the basins. In the Inner Mongolia-Hing'an region, the basins, coming into being mainly in the Cretaceous, are strictly distributed east of the Beishan-Okhotsk shear zone. The Early Cretaceous basins appear mainly in the Greater Hing'an Range and further west and east of the Burya-Jiamusi block, becoming the main coalfields of northeastern

China and small oilfields of Inner Mongolia. The Middle-Late Cretaceous basins occur mainly east of the Greater Hing'an Range, represented by the Songliao basin in which the Daqing Oilfield was formed. The Late Cretaceous-Paleogene basins are present east of the Songliao basin such as the Sanjiang basin. In South China and the Yangtze region, basins were formed mainly in the Middle Cretaceous-Paleogene. The basins in South China and the southern margin of the Yangtze paraplatform are small-sized and are mainly Cretaceous in age. They began to shrink in the Paleogene. Basins in the northern Yangtze paraplatform are large- and medium-sized. They continued to develop in the Tertiary and formed important oilfields. The basins in the Bohai Bay region are mainly Tertiary in age and are obviously different in nature from those in the Inner Mongolia-Hing'an, South China, Yangtze regions. They show clearly the feature of a rift basin characterized by the upwelling of the upper mantle material and the widespread eruption of basaltic magma. The basins in Inner Mongolia and South China manifest post-orogeny extension or strike slipping-extension, whereas the Songliao basin based on the Jilin-Heilongjiang block and the Jianghan and North Jiangsu basins based on the Yangtze paraplatform come in between the last two types in nature. Under these basins the upper mantle swells up but much weaker than that under the Bohai Bay basin.

This is to say that the rift basin system in the continent of eastern China does not strike NNE as previously supposed [24, 25]. In the area east of the Taihangshan Mts. and Greater Hing'an Range, although there occur relatively large-scale Tertiary or Cretaceous extensional faults, this does not necessarily mean that the Greater Hing'an-Taihangshan-Wulingshan line and the gravitational gradient belt with the same name that divide the modern geomorphology of eastern China into two parts in the east and in the west had already come into being by that time. Residual basins of the Cretaceous-Tertiary have been discovered in the present mountainous region in both the Wulingshan and eastern Qinling Mts. The Cretaceous or Tertiary basins in western Henan and western Liaoning transect the modern gravitational gradient belt. Thus it is seen that the unified NNE striking basin-range system is younger. It was formed neither in the Cretaceous nor in the Paleogene, but very probably in the Neogene at nearly the same time with the occurring of the present-day trench-arc system of the West Pacific. Of course, studies in more detail are needed to elaborate this problem.

TECTONIC DYNAMIC REGIME AND ITS EVOLUTION

Here we will mainly discuss the major changes of the dynamic regime in the Phanerozoic, especially in the Mesozoic, in eastern China.

In the Paleozoic, eastern China was mainly under the control of the dynamic system of the Paleo-Asian ocean. The tectonic framework is shown in Figure 2. The continental blocks and oceans generally struck E-W and the main stress of oceanic consumption, continent-continent collision and imbrication was nearly N-S, resulting in the nearly E-W tectonic belts (The direction is referred to the present. The direction in the Paleozoic should be corrected with paleomagnetic data.)

It was believed that eastern China was only controlled by the Pacific system in the Mesozoic. Now there are two stages to be distinguished: in the early stage (Triassic-Jurassic) it was under the joint control of the Tethys and Paleo-Pacific dynamic systems and in the late one, under the control of the Pacific dynamic system with the influence of the formation of the Indian ocean and the closing of the Tethys. The Early Mesozoic tectonic framework of eastern China is shown in Figure 3.

The Tethys was an ancient ocean between Laurasia and Gondwana. It was best developed and its records are best preserved in southwestern China. The Permian-Triassic seaways in the Qinling and South China were both branches of the Tethys. Since the oceanic crust remains (ophiolite) of the

Qinling Tethys are found only in the Anyemaqen Mts. and Triassic turbidite is seen in Fengxian (about 107° E), the eastern Qinling region then was actually an aulacogen of the Tethys, which was situated between the Yangtze and Sino-Korean continents. Turbidite series is only locally developed in the Permian-Triassic of South China and there is no ophiolite. Therefore, South China was also an aulacogen connected to the Tethys. The Paleo-Pacific was an extension of the Tethys.

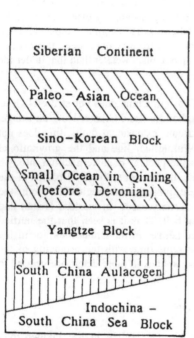

Figure 2. Paleozoic tectonic framework

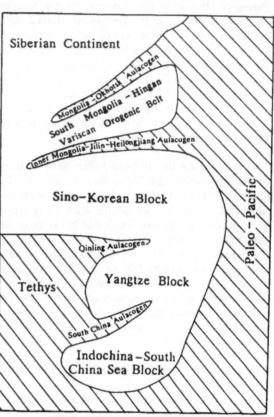

Figure 3. Early Mesozoic tectonic framework

It ran from the Indonesian Archipelago toward Japan, Sakhalin Island and northeastern Russia and was connected to a Mesozoic ocean of western North America. The Mongolia-Okhotsk and Inner Mongolia-Jilin-Heilongjiang belts were both aulacogen-type rifts from the Paleo-Pacific cutting into the Asian continent. In the Middle-Late Triassic, South China, Qinling and Inner Mongolia-Jilin-Heilongjiang were all folded and uplifted and the seas there disappeared, forming nearly E-W or E-W→NE striking tectonic belts. The main stress of the tectonism was still almost N-S. After the Middle Jurassic marine water retreated basically from Mongolia-Okhotsk. The Early Yanshanian movement in the Middle Jurassic completed eventually the dynamic integration of the blocks of eastern China, the Sino-Korean, Yangtze and Indochina-South China Sea blocks, whose apparent paleomagnetic polar wander paths finally came together [7]. The Middle Yanshanian movement in the Late Jurassic-initial Cretaceous (155-140Ma) caused a series of events, the final closing of the Paleo-Pacific and its branch, the Mongolia-Okhotsk aulacogen, the final welding of the Siberian continent and the continent of eastern China, the collision of the landmass of West

Pacifica with the Asian continent, the climax of compression between them, the formation of the great Yanshanian orogenic belt of eastern Asia and the huge epicontinental reactivated belt of eastern China.

As the Mesozoic tectono-magmatic reactivation in eastern China was not the result only of the closing of the Paleo-Pacific of Japan to the east but of the joint action of the Mongolia-Okhotsk aulacogen and the Paleo-Pacific of Japan, and intense compression came from both Siberia and West Pacifica, the eastern and western belts of the epicontinental activated belt of eastern China began to be reactivated when the Paleo-Pacific started to consume (i.e. the Late Triassic) and the structural lines turned to strike NE, whereas the central belt did not become reactivated and the E-W striking structure did not turn to NE-NNE striking until the Late Jurassic-initial Cretaceous whhen the compression reached its climax. Therefore the transformation of the Paleozoic E-W striking tectonic regime of eastern China into the NE-NNE striking one began in the Indosinian and was completed finally in the Yanshanian orogeny which is an overwhelming tectonic movement in eastern China in the Phanerozoic. In the meantime, since the collisional orogeny between West Pacifica and the Asian continent and the imbrication orogeny between the Siberian continent and the Chinese continent both acted obliquely on the continent of eastern China, the tectono-magmatic belts of the Mesozoic epicontinental activated belt of eastern China are usually en enclon in the direction of NE-NNE.

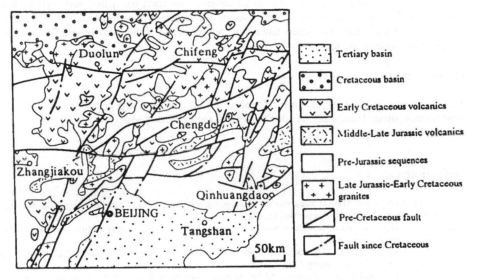

Figure 4. Structural map of South Inner Mongolia – North Hebei Region

The Inner Mongolia and Yanshan-Liaoning regions are the best example which demonstrates that it was the Yanshanian orogeny that resulted in this profound transformation of the tectonic regime of eastern China (Figure 4). There the nearly E-W striking volcanic rocks of the Middle-Late Jurassic Diaojishan Formation (or the Haifanggou Formation) and the red beds of the Tuchengzi Formation (or the Houcheng Formation) represent the sedimentation after the orogeny in the Inner Mongolia-

12

Jilin-Heilongjiang orogenic belt. They are covered at a large angle by the NE-NNE striking volcanic rocks of the Early Cretaceous Zhangjiakou Formation and the Yixian Formation. The fact that the nearly E-W striking Yingshan-Yanshan thrusting system (e.g. the Daqingshan fault, the Chongli-Chicheng fault, the northern marginal fault of Sino-Korea, etc.) was transected at a large angle by the NE-NNE thrusting-folding system also marks the transformation. It is in this tectonic dynamic background that the overpass structure of crust-upper mantle in eastern China has been formed [33].

Because the post-orogenic extension in the Inner Mongolia-Jilin-Heilongjiang orogenic belt occurred in the Late Jurassic (155Ma) [43], the volcanic rocks of the Zhangjiakou Formation base have yielded an isotopic age of 137Ma[26] and the main Sakawa orogeny of Japan occurred in the end-Jurassic-initial Cretaceous, with the post-orogenic strike-slipping starting in the initial Cretaceous (Hauterian) [20], we can say that the profound transformation of the tectonic dynamics of eastern China took place in the end-Jurassic-initial Cretaceous, i.e., c. 155-140Ma.

After the closing of the Paleo-Pacific, a new stage of tectonic evolution began in eastern Asia. The breakup and submergence of West Pacifica and the formation of the modern Pacific ocean resulted in the large-scale rifting of the continental margin of eastern Asia, producing a new oceanic margin and the rift basin system of eastern Asia. At the same time, the ever spreading Indian ocean caused the final closing of the Tethys and the collision of Eurasia with Indian Gondwana, which reworked the tectonics of western China and had an influence on the tectonics of eastern China. For example, the well-known Fenwei rift system is a strike slipping-pull apart basin that may well be formed under the influence of the collision. Thus we say the eastern China has mainly been under the control of the Pacific dynamic system since the Cretaceous and at the same time influenced by the Indian ocean dynamic system.

Acknowledgments

We are thankful to Ma Zongjin, Zheng Yadong, Liu Guodong, Li Jinyi and Min Longrui for their fruitful discussions. Thanks also are due to Dong Xiaoqing, Zhang Miao and Song Yingnian for the drawing of the textfigures.

REFERENCES

1. Bureau of Geology and Mineral Resources of Heilongjiang Province, 1993, Regional Geology of Heilongjiang Province, Geological Publishing House, Beijing, China.
2. Bureau of Geology and Mineral Resources of Jilin Province, 1988, Regional Geology of Jilin Province, Geological Publishing House, Beijing, China.
3. Bureau of Geology and Mineral Resources of Liaoning Province, 1989, Regional Geology of Liaoning Province, Geological Publishing House, Beijing, China.
4. Bureau of Geology and Mineral Resources of Nei Mongol Autonomous Region, 1991, Regional Geology of Nei Mongol Autonomous Region, Geological Publishing House, Beijing, China.
5. Charvet, J., Faure, M., Caridroit, M. and Guidi, A., 1985, Some tectonic and tectogenetic aspects of SW Japan: an Alpine-type orogen in an island-arc position, In: Nasu, N. et al. (eds.), Formation of Active Ocean Margin, Terra Scientific Pub. Co., Tokyo, 791-817.
6. Chen, J. and Dong, J., 1981, A discovery of the Early Triassic conchostrata and its significance in Jilin Province, Chinese Science Bulletin, V. 26, No. 18, 1128-1130.
7. Cheng Guoliang, Sun Yukang, Sun Qingge and Wang Lihong, 1995, Paleomagnetic research on Phanerozoic tectonic evolution of China, Seismology and Geology, V. 17, No. 1, 69-78.
8. Choi, D. R., Vasiliyev, B. I. and Bhat, M. I., 1992, Paleoland, crustal structure and composition under the northwestern Pacific Ocean, In: Chatterjee, S. et al. (eds.): New Concepts in Global Tectonics, Lubbock, 179-191.
9. Dickins, J. M., Choi, D. R. and Yeates, A. N., 1992, Past distribution of oceans and continents, In: Chatterjee, S. et al. (eds.), New Concepts in Global Tectonics, Lubbock, 193-199.

10. Faure, M., Cardidroit, M. and Charvet, J., 1986, The Late Jurassic oblique collisional orogen of SW Japan, new structural data and synthesis tectonics, V. 5, No. 7, 1089-1114.

11. Feng, Q. L., Du, Y. S., Zhang, Z. H. and Zeng , X. Y., 1994, Early Triassic radiolarian fauna of Tongbai region in Henan and its geological significance, Earth Science, V. 19, No. 6, 787-794.

12. Grabau, A. W., 1924, Stratigraphy of China, Pt. 1, Paleozoic and Older, 528.

13. Guo Zhengwu, Deng Kangling, Han Yonghui and others, 1996, The formation and development of Sichuan basin, Geological Publishing House, Beijing, China, 200.

14. He, Z. J., Liu, S. W., Wang, Y. and Ren, J. S., Late Permian-Early Triassic sediments of the Linxi region, Inner Mongolia: evolution and tectonic setting, Regional Geology of China (in press).

15. Helde, T. W., Uyeda, S. and Kroenke, L., 1977, Evolution of the western Pacific and its margin, Tectonophysics, V. 38, No. 1-2.

16. Hsu, K. J., 1981, Thin-skinned plate-tectonic model for collision-type orogeneses, Scientia Sinica, V. 24, No. 1, 100-110.

17. Huang, T. K., 1978, An outline of the Tectonic Characteristics of China, Eclogae Geologicae Helvetiae, V. 71, No. 3, 611-635.

18. Huang, T. K., 1945, On major tectonic forms of China, Geol. Memoirs of National Geological Survey of China, Ser. A, No. 20, 212.

19. Huang, T. K., 1960, The main characteristics of the structure of China: Preliminary conclusions, Acta Geologica Sinica, V. 40, No. 1, 1-37.

20. Ichikawa, K., Mizutani, S., Hara, I., Hada, S. and Yao, A., 1990, Pre-Cretaceous terranes of Japan, Publication of IGCP 224: Pre-Jurassic Evolution of Eastern Asia, Osaka, 413.

21. Institute of Geology, C.A.G.S. (Ren Jishun and Zhang Zhengkun), 1972, Illustrations of China's Geological Map, in: CAGS (eds.): Geological Atlas of the People's Republic of China.

22. Institute of Geology, CAGS, and Wuhan College of Geology, 1985, Atlas of the Palaeogeography of China, Cartographic Publishing House, Beijing, China.

23. Karpinsky, A. P., All-Union order of Lenin Geological Research Institute (VSEGEI), 1983, Geological map of the USSR and adjoining water-covered areas (1:2,500,000).

24. Lee, J. S., 1939, The Geology of China, Murby, London, 528.

25. Lee, J. S., 1973, Crustal structure and crustal movement, Sci. Sin., V. 16, No. 4, 519-559.

26. Li, X., Su, D. Y., Li, Y. G. and Yu, J. X., 1994, The chronostratigraphic status of the Lycoptera-bearing bed, Acta Geologica Sinica, V. 68, No. 1, 87-100.

27. Liu, B. J., Xu, X. S., Pan, X. N., Huang, H. Q. and Xu, Q., 1993, Sedimentary crust of the South China paleocontinent: evolutions and mineralizations, Science Press, Beijing, China, 236.

28. Liu, D. Y., Nutman, A. T., Compston, W., Wu, J. S. and Shen, Q. H., 1992, Remnants of ·3800Ma crust in Chinese part of the Sino-Korean craton, Geology, V.20, 339-342.

29. Ma Xingyuan, 1989, Lithospheric dynamics atlas of China, China Cartographic Publishing House, Beijing, China.

30. Minato, M., Hunahashi, M. et al., 1985, Crustal structure of the Japanese Islands, Japan Sea, coastal part of western Pacific and Philippine Sea, Bulletin of the Japan Sea Research Institute, Kanazawa University, No. 17, 13-42.

31. Mizutani, S. and Yao, A., 1992, Radiolarians and terranes, Mesozoic Geology of Japan, Episodes, V. 14, No. 3, 213-216.

32. Ren Jishun, 1989, Some new ideas on tectonic evolution of eastern China and adjacent areas, Regional Geology of China, No. 4, 1989, 289-300.

33. Ren Jishun, Chen Tingyu, Niu Baogui, Liu Zhigang and Liu Fengren, 1990, Tectonic evolution of continental lithosphere and metallogeny in eastern China and adjacent areas, Geological Publishing House, Beijing, China, 205.

34. Ren Jishun, Jiang Chunfa, Zhang Zhengkun and Qin Deyu, 1980, Geotectonic Evolution of China, Science Press, Beijing, China, 124.

35. Ren Jishun, 1993, relation of east Gondwana to the evolution of the Asian continent (edited by J. J. Pereira, T. F. Ng and T. T. Khoo) Gondwana Dispersion and Asian Accretion, IGCP Project 321, Third International Symposium, Kuala Lumpur.

36. Ren Jishun (Jen Chi-shun), 1964, Preliminary study on some pre-Devonian geotectonic problems of southeastern China, Acta Geol. Sin., V. 44, No. 4, 418-431.

37. Ren Jishun, Chen Tingyu and Liu Zhigang, 1984, Some problems on the division of tectonic units in eastern China, Geological Review, V. 30, No. 4, 382-385.

38. Shao, J. A., Wang, C. Y., Tang, K. D. and Zhang, Q. Y., 1990, Relationship between strata and terrane of the Nadanhada Range, Journal of Stratigraphy, V. 14, No. 4, 286-291.

39. Wang Hongzhen and Mo Xuanxue, 1995, An outline of the tectonic evolution of China, Episodes, V. 18, Nos. 1-2, 6-17.

40. Wu, J. S., Geng, Y. S., Shen, Q. H., Liu, D. Y., Bi, Z. L. and Zhao, B. M., 1991, Significant geological events of Early Precambrian in the North China platform, Geological Publishing House, Beijing, China.

41. Xue Xiangxu, Zhang Yunxiang, Bi Yan, Yue Leping and Chen Danling, 1996, The development and environmental changes of the intermountane basins in the eastern part of Qinling Mountains, Geological Publishing House, Beijing, China, 181.

42. Zhang, G. W., Meng, Q. R. and Lai, S. C., 1995, Tectonics and structure of Qinling orogenic belt, Science in China (B), V. 38, No. 11, 1379-1394.

43. Zheng, Y., Zhang, Q., Wang, Y., Liu, R. and Wang, S. G., 1996, Great Jurassic thrust sheet in Beishan (North Mountains)— Gobi areas of China and southern Mongolia, J. Structural Geology, V. 18, No. 9, 1111-1126.

44. Karpinsky, A. P., All-Union Order of Lenin Geological Research Institute (VSEGEI), 1992. Brief explanatory note on the Geological Map of Russia and Adjacent Areas (1:5 000 000), St. Petersburg.

Proc. 30ᵗʰ Int'l. Geol. Congr., Vol. 6, pp. 15-36
Xiao Xuchang and Liu Hefu (Eds)
© VSP 1997

Formation and Evolution of Cenozoic Continental Rift Belts in East Asia

CUI SHENGQIN

The Institute of Geomechanics, CAGS, 11 Minzu Xieyuan Nanlu, Haidian District, Beijing, 100081, China

Abstract

The peri-Pacific Cenozoic rift domain of the East Asia consisting of four gigantic rift belts parallel to the West Pacific trench-island arc system, is important component of the global continental rifting. These rift belts widely distributed from the continental margin to the intracontinental area and developed since Paleogene or Neogene, including the West Pacific marginal sea rift belt, the Yellow Sea-East China Sea-northern South China Sea continental shelf rift belt, the intracontinental rift belts of the outer part and the inner part of the East Asia. Each of the rift belts consists of several rift systems, and all of the rift belts constitute a unified peri-Pacific Cenozoic rift domain of the East Asia. The West Pacific marginal sea rift belt belongs to the continental marginal rift belt, and the other three rift belts founded on the continental crust belong to the continental or intracontinental rift belts. The geodynamic causes of the three continental or intracontinental rift belts are mainly connected with the super-regional tectonic stress field, the intensive mantle doming, the crust necking and the gravitational adjustment.

Keywords: continental rift belts, formation, evolution, Cenozoic, East Asia

INTRODUCTION

Investigation of rifts and riftogenesis are frontier projects focussed the attention of many geologists and geophysists for long period until present time. The continental rifts occupy special position for studying the lithosphere and its tectonic evolution as one of the major windows for the exploration of deep interior processes [2,16,19,22].

Many different classifications about rifts were provided from different point of views. Here rifts are divided into four kinds according to their crustal types: the continental or intracontinental rifts, the intercontinental rifts, the continental marginal rifts and the oceanic rifts etc. Each kind of rifts can be subdivided into different ranks according to the scale. The basin formed by riftogenesis is called rift basin, it is often consisted of uplifts and depressions. Rift basins together with the accompanied uplifts and horsts between the basins constitute a rift system. Rift systems of similar geological-geophysical features and similar trending directions form a belt of rift systems or a rift belt. And a number of rift belts, formed in similar periods, possessed certain links in their formation and evolution and trending in parallel, constitute an unified rift domain.

The peri-Pacific Cenozoic rift belts and the rift domain of the East Asia are important components of the global continental rifts. These gigantic rift belts parallel to the West Pacific trench-island arc belt and distribute from ocean to continent, from continental margin to intracontinent, and occurred since Late Mesozoic, especially since Early Tertiary or Neogene. They are the West Pacific marginal sea rift belt, the Yellow Sea--East China Sea-northern South China Sea continental shelf rift belt, the intracontinental rift belt of outer part of the East Asia and the intracontinental rift belt of inner part of East Asia. Each of the rift belts is composed of a series of rift basins and several rift systems, and all of the four rift belts constitute an unified peri-Pacific Cenozoic rift domain of East Asia.

The four rift belts developed on different types of the crust respectively. The west Pacific marginal sea rift belt developed on the continent-ocean transitional crust in the east of this area belongs to the continental marginal rift belt and will not be discussed in detail in this paper. The other three rift belts founded on continental crust belonging to the continental or intracontinental rifts will be discussed in detail as the main topics of this paper.

THE GEOLOGICAL-GEOPHYSICAL FEATURES OF THE CENOZOIC CONTINENTAL RIFT BELTS IN EAST ASIA

Continental Marginal Rift Belt

In the transitional zone between the continental and the oceanic crust of the eastern margin of the East Asian continent developed the west Pacific trench-island arc belt.

17

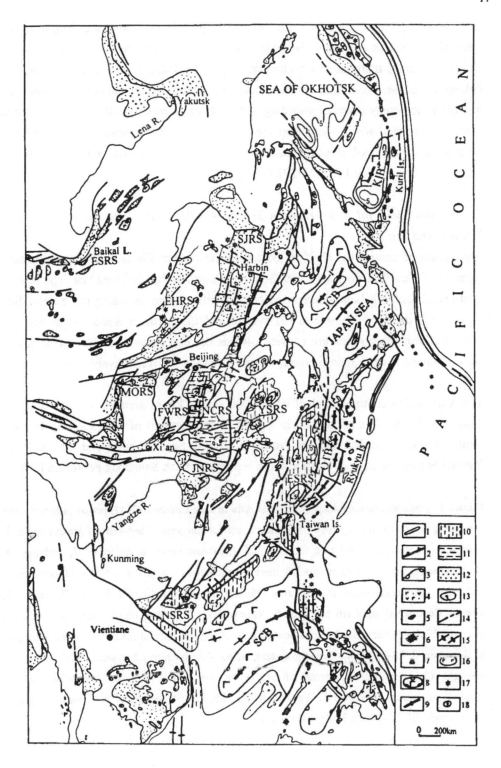

In its adjacent west formed the west Pacific marginal sea rift belt extending from the Okhotsk Sea-the Sea of Japan-the Okinawa trough to central and south part of South China Sea. In the marginal sea rift belt developed the double-concave crustal structure belonging to the transitional crust, where the Moho below deep-water basins or troughs is obviously convex upwarding, the continental crust was thinned, the oceanic crust was accreted and in some deep sea basins formed new oceanic crust. In the west of this belt developed three gigantic continental rift belts within the East Asian continental crust [3,5,9,11,16,27-29](Fig.1,2).

Continental Shelf Rift Belt From the Yellow Sea-the East China Sea-the Northern South China Sea

The region of continental shelf is located in the eastern China continent, extending from the Yellow Sea-the East China Sea to the northern South China Sea. The related continental shelf rift belt is composed of three rift systems including the Yellow Sea rift system, the East China Sea rift system and the northern South China Sea rift system. These rift systems are convex towards the east generally. Their main basins are distributed in dextral en echolon generally trend in NE respectively (Fig.1,2).

The Yellow Sea continental shelf is 380,000 square km in area. Its average depth is 44m. The Yellow Sea rift system is located in the east part of the Tancheng-Lujiang fault zone. It is divided into the Northern Yellow Sea rift basin and the Southern Yellow Sea rift basin from north to south seperated by the Shandong Peninsula [30].

Figure 1. Schematic tectonic map of the East Asia in Late Cenozoic (N-Q) period (adapted from Cui Shengqin and Li Jinrong, 1990). 1, trench. 2, subduction zone. 3, boundary of oceanic basin. 4, marine sedimentary and volcanic formation. 5, ultrabasic rocks. 6, ophiolite. 7, melange. 8, marginal sea basin. 9, spreading axis. 10, marine sediments. 11, paralic sediments. 12, continental sediments. 13, isopach (km). 14, fault zone. 15, anticline and syncline. 16, basalt. 17, crater. 18, granite. Continental shelf rift belt: YSRS--Yellow Sea rift system, ESRS--East China Sea rift system, NSRS--Northern South China Sea rift system. Intracontinental rift system of the outer part of East Asia: SJRS--Songliao-Jieya rift system, NCRF--North China rift system, JNRS--Jianghan-Nanyang rift system. Intracontinental rift belt of the inner part of East Asia: ESRS--East Siberia (Baikal) rift system, FWRS--Fen-Wei rift system, MORS--rift system of north-west margin of Ordos block, EHRS--Erlian-Hailar rift system.

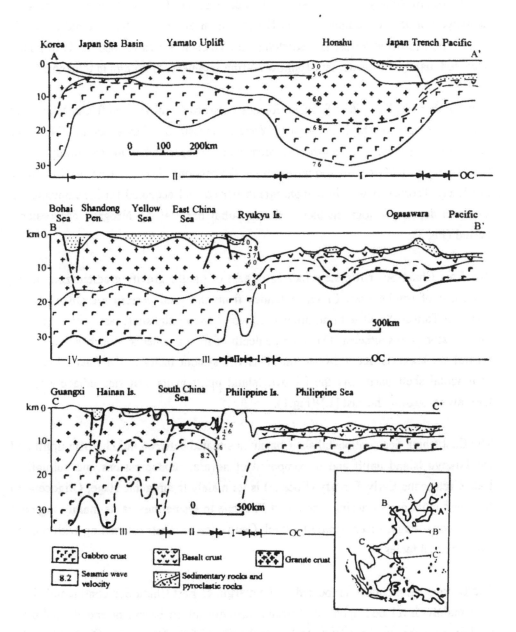

Figure 2. Sectional sketch map of the crustal structure in the East Asian Peri-Pacific region (after the Second Marine Survey Brigade, MGMR, with supplementary data). OC--West Pacific oceanic crust region, I--West Pacific trench-island arc belt, II--West Pacific marginal sea rift belt, III--Yellow Sea-East China Sea-northern South China Sea rift belt, IV--intracontinental rift belt of the outer part of East Asia.

The Northern Yellow Sea rift basin: its two en echelon Mesozoic-Cenozoic basins are developed on the foundation of Sino-Korea craton and on the background of uplifts where the Mesozoic terrestrial sediments and volcanic series formed. In Cenozoic it accepted mainly the Quaternary shallow marine sediments of 2500m in thickness.

The Southern Yellow Sea rift basin: locates between Qianliyan uplift and Wunansha uplift, developed on the foundation of Yangtze massif, and has apparent ENE trend generally. Its southern and northern depressions are separated by the central uplift. the total thickness of Cenozoic sediments in the depressions almost reaches to 4000m. In the Early Tertiary it was in a taphrogenic period and accepted thick sediments. Its southern depression join together with the Subai basin at the Neogene depressional period (Fig.3).

The East China Sea rift system locates in east of the the gravity gradient zone along the coast of southeastern China, extendes from the southwestern margin of Korea Strait to Taiwan Strait with an arcuate convex towards the southeast. The bottom of the sea sloped to southeast. The average depth of its sea water approaches to 72m. Its crustal thickness is about 24-26 km. This rift system includes the East China Sea continental shelf basin and the Diaoyu Island uplift in its east side adjacent to the Okinawa trough in the east [12](Fig.1).

The East China Sea continental shelf basin: is situated between the Zhe-Min uplift and the Diaoyu Island uplift and is composed of several subdepressions and subuplifts. Late Cretaceous-Early Tertiary (Eocene) is its mainly taphrogenic stage, Oligocene to Miocene is its downwarping stage, and Pliocene to Quaternary is its mainly regional depressional stage, accompanied by thick Cenozoic sediments alternating with marine and terrestial facies.

The Diaoyu Island uplift: is located in the margin of East China Sea continental shelf. It is mainly developed upon the Tertiary strongly folded basement and its sediment thickness reaches to about 1000-1500m in the period from Neogene to Quaternary.

The region between east of the Diaoyu Island uplift and the Ryukyu Island uplift is the Okinawa trough. Its sectional bottom looks like U-shape with bigger depth in the south (deepest 2273m) and shallower in the north (nearly 1000m in depth). The heat flow

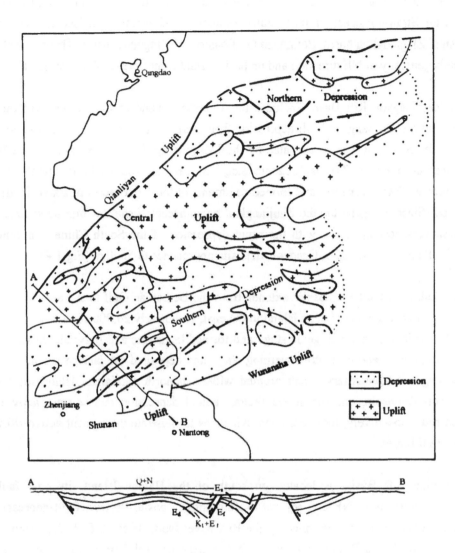

Figure 3. Regional tectonic map and cross section in the southern Yellow Sea rift basin-Subai rift basin (after Li Desheng, 1981, simplified).

values within the trough are 2-4 HFU in general. The trough is one of the back-arc basin of Ryukyu Island arc belt started spreading in Late Miocene and is still active at present on the basis of Tertiary folded metamorphic basement with the Quaternary active sea bottom volcanism in it. The thickness of its crust is 15-17km.

The Northern South China Sea continental shelf rift system locates in the southwest of Taiwan Strait-northeastern Indo-China Peninsula and mainly consists of three rift basins: Zhujiangkou basin, Beibuwan Gulf basin and Yinggehai basin. This rift system trends generally in NE direction and its basins trend from ENE-EW to NW [30].

The Zhujiangkou Rift Basin: includes the Wanshan Island uplift, the northern fault-slop and the Zhujiangkou fault-depression from northwest to southeast with a strike of ENE. Its southeast part is separated by the Dongsha uplift from the South China Sea central deep basin. The taphrogenic stage of Zhujiangkou basin is mainly Late Cretaceous-Early Tertiary, and formed sediments of terrestial facies in Early Tertiary period. Since Neogene the the Zhujiangkou basin accepted shallow marine sediments extensively accompanying with the rapid depression of the South China Sea center basin. Its total thickness of Cenozoic deposits approaches to 5000 m (Fig.1,4).

The Beibuwan Gulf Rift Basin: extends from the Beibuwan Gulf to the area between Leizhou Peninsula and Hainan Island in the east, with nearly 20,000 square km in area. Early Tertiary is its taphrogenic stage and formed lacustrine and fluivatile deposits in this stage. Neogene is its downwarping stage and deposited the marine sediments in this stage. In Quaternary basalt erupted widely in north of Hainan Island-south of Leizhou Peninsula, and the active faults, faulted depression and seismic activities occurred associatively, these are good evidences to illustrate the present active rifting exits in this area.

Yinggehai Rift Basin: is located on south of the Hainan Island. Its east fault-depression trends in NE and adjoins to Zhujiangkou basin. Its west fault-depression trends in NW and was controlled by the Red River fault. Its south fault-depression in east of Indo-China Peninsula trends in NS. So the general framework of the whole Yinggehai rift basin takes the shape of " Y ". More than 10000m Tertiary sediments deposited in the Yinggehai basin and Late Cenozoic basalt erupted extensively in its neighboring southeast Indo-China Peninsula (Fig.5).

Intracontinental Rift Belt in the Outer Part of the East Asian Continent
Since the end of the Mesozoic especially since the Early Tertiary a NEN trending intracontinental rift belt composed of series of rift systems developed in the outer part of the East Asian peri-Pacific region, extending from the southwest Sea of the Okhotsk

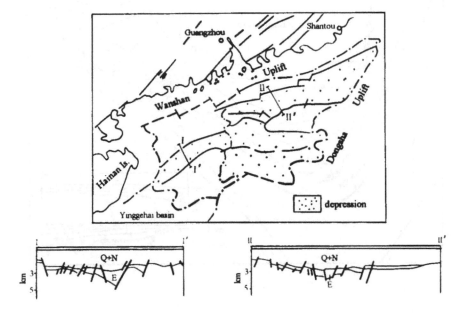

Figure 4. Regional tectonic map and cross sections in the Zhujiangkou rift basin (after Wang Shangwen et al. 1983, simplified).

-Jieya River area to east of the gravity gradient zone of the Greater Khingan Mts.-Taihang Mts.-Wuling Mts. The components of this rift belt include the Songliao-Jieya rift system, the Northern China rift system and the Jianghan-Nanyang rift system from north to south, and extend continuely to the Beibuwan Gulf basin in the south part (Fig.1).

The Songliao-Jieya rift System: consists of the Jieya River basin located in north of Heilongjiang River, Songliao basin and Sanjiang basin etc. Its basement is the epizonal metamorphic and igneous rocks of Late Paleozoic Variscian stage and its crustal thickness is nearly 29-35km. The mainly rifting period of each basin of the Songliao-Jieya rift system is Late Mesozoic. The rifting became withering away in Cenozoic and almost extincted in Late Cenozoic under the influences of the ôpening of the Sea of Japan. Its largest basin -- the Songliao basin (Fig.6) will be discussed in detail as an example.

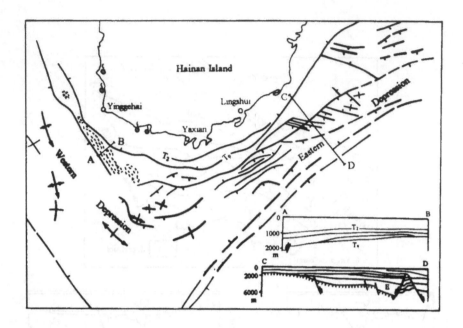

Figure 5. Regional tectonic map and cross section in the Yinggehai rift basin (after Wang Shangwen et al. 1983, simplified).

The Songliao rift basin consisted of ten isolated NNE-trending downfaulted smaller basin groups was mainly controlled by the basement faults in Middle and Late Jurassic stage[26]. The volcanic-sedimentary rocks in the Songliao basin approaches to more than 1000m. Its taphrogenic stage continues to the initial Early Cretaceous followed by downwarping stage when a unified gigantic basin of 260,000 square km formed. The rift basin became withering away since Late Cretaceous. After the Tertiary and Quaternary disintegration and planation the overlying strata of 100-150m thick formed on the top of the basin as the result of withering and equilibrium stage. For the basin at present, the crustal thickness is 30-32km, and the average heat flow value is about 1.9 HFU.

The North China rift system: is the largest Cenozoic rift system of the intracontinental rift belts of East Asia. It is composed of three rift basins: the Xialiaohe-the Northern North China rift basin (Fig.7) and the Southern North China rift basin in the west of Tancheng-Lujiang fault zone, and the Subai rift basin located in the east of Tancheng-Lujiang fault zone.

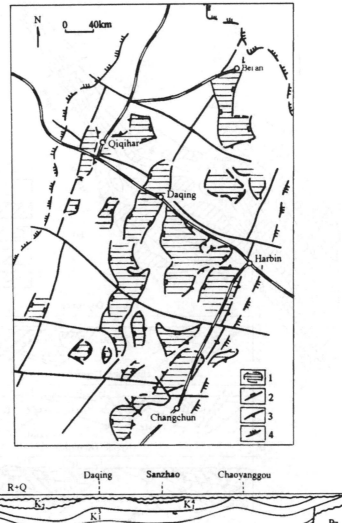

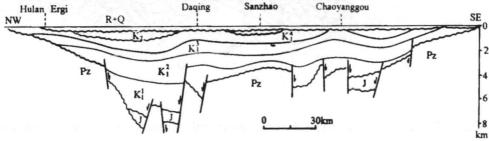

Figure 6. Regional tectonic map and cross section in the Songliao rift basin (after Tian Zaiyi et al. 1992, simplified).

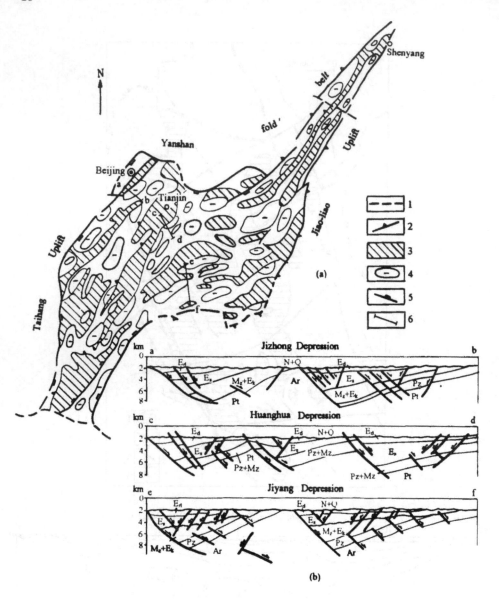

Figure 7. Regional tectonic map and cross sections in the Xialiaohe-northern North China rift basin (after Tian Zaiyi et al. 1991, simplified). 1, boundary with Southern North China rift basin. 2, missing of Paleogene. 3, uplift. 4, sag. 5, fault. 6, line of cross section.

The Xialiaohe-Northern North China rift basin developed in the Sino-Korea massif is mainly consisted of 6 downwarped depressions such as the Xialiaohe River depression, the central Bohai Sea depression, and the Jizhong depression etc., and more than fifty secondary lystric-graben shape fault-depressions were formed. Its total area reaches to about 200,000 square km (Fig.7). In Eogene--its taphrogenic stage, the lacustrines oil source strata system developed in this basin. In Neogene-Quaternary--its downwarping stage, widespread sedimentation occurred in this basin. The total thickness of Cenozoic sediments in the Xialiaohe-Northern North China rift basin is generally 6000-7000m and reaches to 12000m in the Bohai Bay. The crust thinned to 30-40km thick in general, and the thinnest crust of the Bohai Bay is only 28km thick. The low velocity layers within the crust distribute extensively, the Moho appears to be a transitional layer, the asthenosphere upwarded and the lithosphere thinned to 60-80km in this basin. The heat flow value appears to be higher in this region, about 1.6 HFU along the boundary of the North China Plain and more than 2.0 HFU along the coast of Bohai Bay. The Cenozoic mantle-originated basalt erupted in some depressions, and strong seismic activities occur intensively at present, especially in the central and northern margin of the basin [8].

The Southern North China basin is mainly located between the south of Yellow River and the north of Huaihe River in the Sino-Korea massif. In its north it is separated from the Xialiaohe-Northern North China rift basin by the Xinxiang-Lankou fault, and the fault-depression changes trending to WNW. The development feature of this basin is similar to the Xialiaohe-Northern North China basin.

Subai rift basin is located in east of Tancheng-Lujiang fault. It is mainly developed on the foundation of Yangtze massif. After the Jurassic-Cretaceous downwarping stage, this basin entered into the Early Tertiary taphrogenic stage and deposited the dark oil source strata system until the Neogene downwarping stage when the Subai basin connected with the southern depression of Southern Yellow Sea basin (Fig.3).

Jianghan-Nanyang rift system: is a small rift system developed on the Dabie-Qinling orogenic belt and the Yangtze massif in Late Cretaceous-Early Tertiary, including Jianghan basin and Nanyang-Xiangyang basin which are joined together by Hanshui River graben of NW-trending.

Jianghan rift basin on the Yangtze massif is nearly 28,000 square km in area. Its taphrogene stage is Late Cretaceous-Early Tertiary. In Oligocene this basin deposited the oil source strata consisting of clastic rocks, gypsum salt and dark mudstones. Neogene-Quaternary is the downwarping stage in this region and deposited a great deal of sediments of river and locustrine faces in the whole basin.

Nanyang-Xiangyang rift basin developed on the Proterozoic and Paleozoic metamorphic rocks as its basement is a small rift basin with the area of 5,000 square km. Its taphrogenic stage is also Late Cretaceous-Early Tertiary when a huge thickness of dark mudstone mingled with sandstone and conglomerate consisted of oil source stratigraphic system formed in this basin. This innermountainous basin changed from uplifted into rising disintegrational state in Neogene-Quaternary.

Intracontinental Rift Belt in the Inner Part of the East Asian Continent
From northern part of Mongolia and East Siberia to the west of Greater Khingan Mts.-Taihang Mts. gravitational gradient zone of northern China, within the inner part of East Asian continent, a series of terminal Mesozoic-Cenozoic rift systems developed on the general background of high mountains and uplands or plateaus, including the East Siberia (Baikal) rift system, Fen-Wei rift system, northern margin-western margin of Ordos block rift system and Erlian-Hailar rift system etc., constituting an intracontinental rift belt (Fig.1).

The East Siberia (Baikal) rift system: lies in the south part of East Siberia upland, including a long belt winding from north Mogolia, east Sayan-Lake Baikal to the Udokan Range of southern Yakutia, with the general framework appears in "S" shape. The whole rift system is 1800km in length, and contains thirteen rift basins of different sizes such as the south Baikal rift basin, the Tunka rift basin and the north Baikal rift basin etc.(Fig.8).

The early stage of East Siberia rift system (E$_{1-2}$-N$_2$) was marked by less topographic contrast, gentle fault-depression and formation of deposits limited in South Baikal rift basins etc. The mantle-originated basalt erupted in the west and northeast region of the rift system. A series of new rift basins formed accompanying the extension of the descending range in the late stage of the rift system (N$_2$-Q). The upper mantle low velocity layer exists under the rift system where the low velocity anomalous layer and mantle dome uplifted, and formed the crustal-mantle mixed layer of 1000km in length. This rift system is also a telluric heat flow anomalous zone with its heat flow values 2-3 times higher than that of the regular district and an active seismic zone at present time [13-15,23].

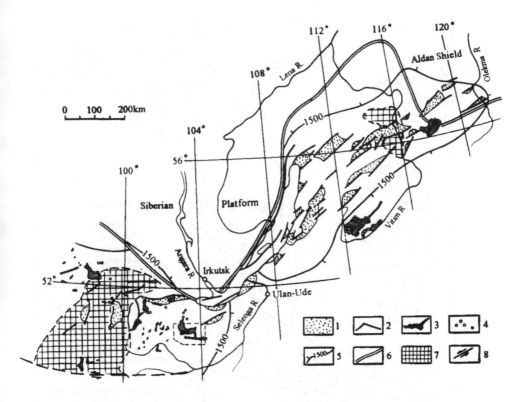

Fiure 8. Schematic map of the general structural and geomorphic setting of the Baikal rift system (after N.A.Logatchev, 1993). 1, sedimentary infill. 2, major faults. 3, volcanic fields. 4, volcanic cones. 5, 1500m contour line of the Cretaceous-Paleogene erosion surface. 6, rim of the Siberian platform. 7, microcontinents. 8, Bolnai strike-slip fault.

The Fen-Wei rift system: is located between the Yinshan-Yanshan and Qinling two latitudinal tectonic belts, and is composed of ten rift basins with different sizes such as the Weihe basin, the Linfen basin and the Datong basin etc. Its general framework appears in "S" shape and and extends 1200km in length (Fig.9).

In the early rifting stage of Fen-Wei rift system (E2-N1) formed the Weihe rift basin and erupted the intracontinental alkalini basalt in the north part of Xinxian rift basin. In the late rifting stage (N2-Q) of Fen-Wei rift system every rift basin depressed rapidly, Quaternary accumulation in the south side of the Weihe basin approached to 2000m, and new basins such as the Taiyuan basin, the Datong basin and the Yanqing basin etc.

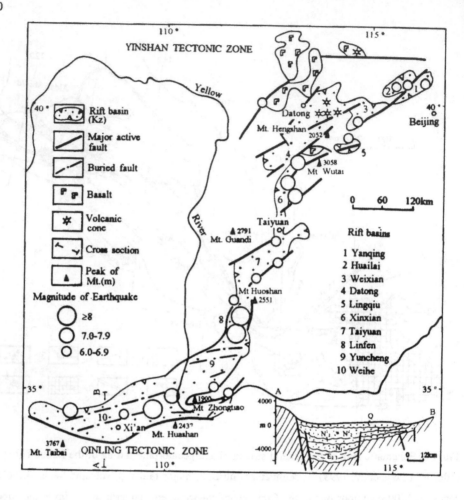

Figure 9. Schematic map of regional tectonics of the Fen-Wei rift system (after Cui Shengqin and Li Jinrong, 1988).

formed in this stage in central and northern segments of this rift system. In Quaternary the intracontinental alkaline basalt effused in Datong basin and its northen border areas. Three thin low velocity layers existed within the crust of Fen-Wei rift system. The crustal thickness of Fen-Wei rift system is 36-38km while the crustal thickness of the neighboring region is more than 38-40km in general. Near the Xian city in the middle of the Weihe basin the crustal thickness is only 34 km. Many hot spring groups distribute along the anormal geothermal zone of this rift zone, and the heat flow values in the southern margin of Linfen basin reach up to 2.13 HFU. Many earth fissures happened in the Xian and Datong areas and the present strong seismic activity occur in the Fen-Wei rift system [4,6].

The north-west margin rift system of Ordos block: includes the E-W trending Hohhot- -Baotou rift basin in front of the Daqingshan Mts. in the north of Ordos block, the N-S trending Yinchuan basin in the west margin of Ordos block and the Linhe-Jilantai basin between them. These rift basins are limited and surround by the Cenozoic faults generally with the Cenozoic sediments of 4000-7500m thick. The Moho uplift extended above to 2-3 km below this rift system. It is also an active tectonic belt and seismic belt at present.

The Erlian-Hailar rift system: displays in arc shape and lies on the north part of Yinshan uplift and the west part of Greater Khingan Mts. It is mainly developed on the basis of Paleozoic folded zone. The Erlian basin with EW-ENE trend is located on south side of this rift system with the Sunite uplift in it. The Hailar basin is located in the north of Erlian basin and trends in NE. Its main rifting period is Late Jurassic-Early Cretaceous when the oil source rocks and coal source strata of deep lacustrine and swamp facies formed in this rift system. The rift basins uplifted and withered away steppedly after the Late Cretaceous and only accepted small and thin Cenozoic sediments. But in Neogene-Quaternary intracontinental mantle basalt erupted in several places of this rift system, illustrating the active deep interior process existed within it.

TECTONIC EVOLUTION AND GENETIC MECHANISM OF THE CENOZOIC CONTINENTAL RIFT BELTS IN EAST ASIA

Tectonic Evolution of the Continental Rift Belts
Certain differences and some similarities exist among the three Cenozoic continental rift belts in East Asia. Separated by the E-W trending Qinling-Dabei and Yinshan tectonic zones, the Cenozoic continental rift belts in the inner and outer East Asian continent can be divided into north, central and south segments of different evolutional features [5](Fig.1).

The rift systems of the north segment mainly include the withering Songliao-Jieya and the Erlian-Hailar rift systems with their taphrogenic stage in Late Mesozoic. These rift systems getting withered away and extincted in Cenozoic under the influences of the opening of the Sea of Japan. The widely erupted Neogeng-Quaternary basalt in these rift systems may be a special form of Cenozoic rifting in this regions.

The rift systems of the central segment include the newly formed North China rift system, the Fen-Wei rift system and the rift system of north-west margin of Ordos block. From terminal Cretaceous to Paleocene uplifting and erosion took place in many places of these rift systems, corresponding to the initial mantle doming.

The rift systems of the south segment were in the compressional status between the Taiwan Island arc and northeast corner of Indian Plate since Late Eocene. At the beginning most of rift systems were in uplifting state except the Jianghan-Nanyang rift system. In Early Tertiary rift basins of this segment always developed on the Cretaceous rift basins, mostly succeeded their precursors.

The Cenozoic rifting of most areas of the Yellow Sea, the East China Sea and the Northern South China Sea rift systems of the continental shelf rift belt usually followed the Late Mesozoic taphrogeny.

In Early Himalayan, from terminal Mesozoic to Early Tertiary, down-faulting took place on the background of general uplifting in the East Asian continent. For most of the rift systems of the three Cenozoic continental rift belts in East Asia, the Paleogene taphrogenic stage might correspond to the period of the intensive regional mantle doming and crust necking. At the same time, the mantle-derived basalt magma erupted in many places within the Xialiaohe, Bohai Bay, northern North China and Subai rift basins. The orogenesis from the terminal Early Tertiary to Early Miocene greatly changed the tectonic framework and environment of the East Asian. These changes were reflected not only by the stratigraphic contact relationship between Upper and Lower Tertiary, but also by the transition of rifting from taphrogenic stage to downwarping stage in the tectonic evolution of the East Asian continental rift belts.

In Late Himalayan, from Miocene or Pliocene, the rift belts in East Asian continent entered into a steadily spreading stage of downwarping and drapping corresponding to the stage of attenuation of mantle doming and gravitational adjustment. Since Late Miocene - Pliocene, the Ochotsk Sea basin, the Japan Sea basin, the Okinawa trough and the central South China Sea basin of the West Pacific marginal sea rift belt entered into largely spreading stage when the Taiwan movement or Island Arc disturbance took place in the island arc belt. The compression, shearing and subduction between continent and ocean became intensified, resulted ultimately the trench-arc-basin system in the East Asian peri-Pacific region (Fig. 1,2).

Genetic Mechanism of Continental Rifting in East Asia

The problem of genetic mechanism and geodynamic causes of the continental rifting is very complicated. Disagreement existed upon the genetic mechanism of Cenozoic continental rift belts in East Asia. In recent years, many geologists analyzed and discussed the genetic mechanism of the Cenozoic East Asian continental rift systems and rift belts from different point of views [1,2,4,5,7,10,13-18,20,21,23,26,31].

After analyzed the slip line field of the East Asia, P. Tapponnier and P. Molnar et al. [21,24,25] suggested that the rift systems in East Asia were resulted from the collision of the Indian Plate with the Eurasian Plate which caused the successive dislocation and pull-apart of different blocks of East Asia along the slip line. They pointed out that the South China Sea basin was first occurred 50 Ma ago, followed by rifting of the Fen-Wei rift system in about 20 Ma and then the formation of the much later Baikal rift system, and the strongest rifting stage in East Asia has not come yet. But their opinion is not comformed with the geological facts. The South China Sea, Fen-Wei and Baikal rift systems are very similar to each other in the formational and evolutional stages, and the riftogenic intensity in different rift belts appears weaker in the west and stronger in the east. it seems that a reasonable genetic mechanism needs to combine closely the data of geology and geophysics, superficial structure and deep interier process, tectonics of intracontinent and continental margin, present structures and historical structures, etc.

The subparallelism and slightly en echolon arrangement of the intracontinental rift belts and the continental-shelf rift belt as well as the marginal sea rift belt in the East Asian per-Pacific region reflect a dominantly extensional stress field existing extensively in the involved areas as the general background of formation and evolution of the East Asian peri-Pacific Cenozoic rift domain. The Cenozoic regional stress field is also marked by alternation of extension and compression both in space and time on the other hand.

To sum up, the genetic mechanism of the Cenozoic continental rift belts in the East Asia should be preliminarily considered as the result of combined actions of continent and ocean, intracontinent and marginal continent, the surface and the deep Earth, tectonic pattern and thermal upwelling or little scale mantle convection under the condition of the rotation of the Earth and the controlling of the super-regional stress field.

Acknowledgements

First of all I would like to express my deep gratitude to Professors Sun Dianqin, Chen Qingxian, Wang Hongzhen, Ma Xingyuan and V.E.Khain for all their helpful suggestion in my research on tectonics in the East Asian peri-Pacific region. I am very grateful to Professors N.A.Logatchev, S.I.Sherman and A.A.Bykxarov for our comparative research of the Baikal and Fen-Wei rift systems, and thank Professor J.Mercier for our cooperative work in the Weihe area. I also especially acknowledge Professors E.E.Milanovsky, E.V.Artyushkov, Tian Zaiyi, Liu Guangding, Ma Zongjin and Liu Hefu for useful discussions. Besides, I would like to thank my colleaques Li Jinrong, Wu Zhenghan, Feng Xiangyang, Ren Xifei and Shang Ling for their help in the presentation of this manuscript and figures.

REFERENCES

1. E.V.Artyushkov, F.A.Letnikov and V.V.Ruzhich. The mechanism of formation of the Baikal basin, *Journal of Geodynamics*. 11, 277-291 (1990).

2. K.Burke. Intra-continental rifts, *Continental Tectonics* (1980).

3. Cui Shengqin, Li Jinrong and Zhao Yue. On the Yanshanian movement of peri-Pacific tectonic belt in China and its adjacent areas,in:*Scientific Papers on Geology for International Exchange Prepared for the 27th Int'l.Geol.Conr.* pp.221-234. Geological Publishing House (1985).

4. Cui Shengqin and Li Jinrong. Comparative tectonic analysis of the Fen -Wei graben and Baikal rift system, *Bullitin of Chinese Academy of Geological Sciences*. 15,25-37 (1988).

5. Cui Shengqin and Li Jinrong.Himalayan tectonic evolution in the East Asian peri-Pacific region, *Acta Geologica Sinica*. 3 (3), 233-246 (1990).

6. Cui Shengqin and N.A.Logatchev. Similarities and differences in Fen - Wei and the Baikal rift system, *Abs. 29th Int'l.Geol.Congr.* Vol.1 of 3(1992).

7. Deng Jinfu, Zhao Hailing, Mo Xuanxue,Wu Zongxu and Luo Zhaohua.*Continental roots-plume tectonics of China--key to the continental dynamics,* Geological Publishing House (1996).

8. Ding Guoyu (Chief Editor). *Lithosphere dynamics of China,*Seismological press (1991).

9. T.W.C.Hilde, S.Uyeda and L.Kroenke. Evolution of the western Pacific and its margin, *Tectonophysics.* 38, 145-165 (1977).

10.V.E.Khain. Origin of the central Asian mountain belt: collision or mantle diapirism, *Journal of Geodynamics.* 11, 389-394 (1990).

11.V.E.Khain. *Geology of Northern Eurasia,* Berling: Gebruder Borntraeger (1993).

12.Liu Guangding. Geophysical and geological exploration and hydrocarbon prospects of the East China Sea, *China Earth Science.* 1(1), 43-58 (1989).

13.N.A.Logatchev. The Baikal rift system, *Episodes,*7(1), 38-42 (1984).

14.N.A.Logatchev. History and geodynamics of the Lake Baikal rift in the context of the Eastern Siberia rift system: a review, *Bull.Centres Rech.Explor.-Prod.Elf Aquitane.* 17(2), 353-370 (1 993).

15.S.V.Lysak. Terrestial heat flow of continental rifts, *Tectonophysics.* 143, 31-41 (1987).

16.Ma Xingyuan, Liu Hefu, Wang Weixiang and Wuang Yipeng. Meso-Cenozoic taphrogeny and extensional tectonics in Eastern China, *Acta Geologica Sinica.* 57 (1), 22-32 (1983).

17.Ma Xingyuan and Wu Daning. Cenozoic extensional tectonics in China, *Tectonophysics.* 133, 243-255 (1987).

18.Ma Zongjin and Du Pinren. *The problems on recent crustal movement,* Geological Publishing House (1995).

19.E.E.Milanovsky. Some problems of rifting development in the Earth's history,in:*Tectonics and geophysics of continental rifts.* I.B.Ramberg and E.R.Neumann eds. pp.385-399. Reidel.Dordrecht (1978).

20.E.E.Milanovsky. Main stages of rifting on the territory of China, *Results of researches on the international geophysical projects.* Moscow (1991).

21.P.Molnar and P.Tapponnier. Cenozoic tectonics of Asia: effects on the continental collision, *Science.*189, 418-426 (1975).

22.P.Morgan and B.H.Baker (eds). *Processes of continental rifting,* Els.Sci.Pub. (1983).

23.S.I.Sherman. Faults and tectonic stresses of the Baikal rift zone, *Tectonophysics.* 208. 297-307 (1992).

24.P.Tapponnier and P.Molnar. Slip-line field theory and large-scale continental tectonics,*Nature.* 284 (5584), 318-324 (1976).

25.P.Tapponnier,G.Peltzer,A.Y.Ledain,R.Armijo and P.Cobbold. Propogating extension tectonics in Asia: new insight simple experiments with plaoticene, *Geology.* 10, 611-616 (1982).

26. Tian Zaiyi, Han Ping and Xu Keding. The Mesozoic-Cenozoic East China rift system, *Tectonophysics*. 208, 341-363 (1992).

27. Tong Chongguang. Some characteristics of petroleum geology of the rift system in the Eastern China, *Petroleum Journal*. 3 (1980).

28. Wang Hongzhen, Yang Sennan and Li Sitian. Mesozoic and Cenozoic basin formation in East China and adjacent regions and development of the continental margin, *Acta Geologica Sinica*. 57 (3), 213-223 (1983).

29. Wang Hongzhen and Mo Xuanxue. An outline of tectonic evolution of China, *Episodes*. 18, 6-16 (1995).

30. Wang Shangwen et al. *Petroleum Geology*, The Oil Industry Publishing House (1983).

31. Zhang Wenyou et al. *Marine and continental geotectonics of China and its environs*, Science Press (1986).

Proc. 30ᵗʰ Int'l. Geol. Congr., Vol. 6, pp. 37-46
Xiao Xuchang and Liu Hefu (Eds)
© VSP 1997

Structural Characteristics of Meso-Cenozoic Continental Rifts in Eastern China

CHEN FAJING
Department of Geology and Energy Resources, China University of Geology, Beijing, 100083, China
QI JIAFU
Department of Earth Sciences, University of Petroleum, Beijing, 102200, China
CHEN ZHAONIAN
Department of Geology and Energy Resources, China University of Geology, Beeijing, 100083, China

Abstract

The Meso-Cenozoic continental rifts were formed in eastern China during early Jurassic, early Cretaceous and Palaeogene, and evolved into a widespread continental sag during middle Jurassic, late Cretaceous and Neogene-Quarternary. Taking the Songliao and Bohai Bay basin provinces as examples, the structural characteristics are discussed in this paper, namely: (1) structural style; (2) transfer zone; (3) Basin modification; (4) subsidence and thermal regime and (5) volcanic rocks. In this paper, we especially point out that there were two relatively independent but interrelated Cenozoic tectonic systems (extensional and strike-slip tectonics) developing in Bohai Bay basin province. Taking all of these points into account and based on the regional tectonic setting, two different geodynamic models of basin formation are proposed.

Keywords: strctural characteristics, extensional tectonics, strike-slip tectonics, geodynamic models of basin formation.

1. Temporal and spatial aspects of the continental rifts

The Meso-Cenozoic continental rifts were formed during early Jurassic, early Cretaceous and Palaeogene in eastern China. As for these three rift systems, the occurrence of early Jurassic rifts can't be easily established, because they were destroyed at the end of middle Jurassic by horizontal compression.

The distribution of early Cretaceous rifts in northeastern China is shown in Fig 1. They are distributed in three zones. The Hailer and Erlian basin provinces are located in western zone and are distributed in NE trending. The Songliao basin province is distributed in central zone and trends in NNE direction . Only Sanjiang-Jixi-Boli-Yanji basin province located in eastern zone is relict basins and is distributed in NS, NNE, EW and NEE trending.

The Bohai Bay basin province, Which includes 3 geographic units, namely northern North China plain, sea area of Bohai Bay and Xialiaohe plain, is a very important petroleum basin in eastern China (Fig. 2). Its configuration looks like a capital letter (N) with an expanding in the middle part, the basin

38

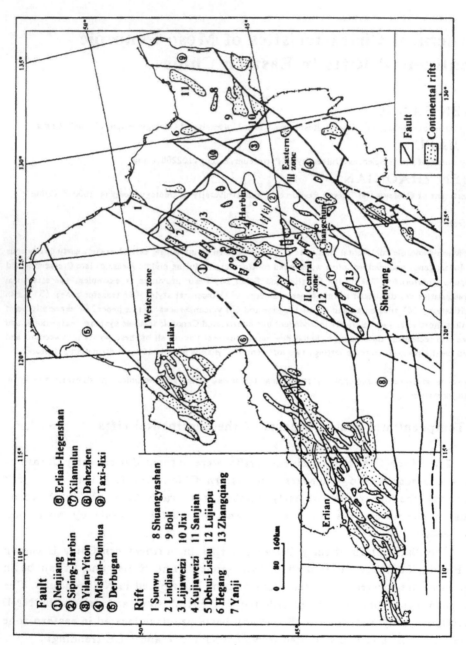

Fault

① Nenjiang ⑥ Erlian-Hegenshan
② Siping-Harbin ⑦ Xilamulun
③ Yilan-Yiton ⑧ Dahezhen
④ Mishan-Dunhua ⑨ Taxi-Jixi
⑤ Derbugan

Rift

1 Sunwu 8 Shuangyashan
2 Lindian 9 Boli
3 Lijiaweizi 10 Jixi
4 Xujiaweizi 11 Sanjian
5 Dehui-Lishu 12 Lujiapu
6 Hegang 13 Zhangqiang
7 Yanji

0   80   160km

⬛ Fault

▨ Continental rifts

Fig. 1 Distribution of early Cretaceous continental rifts in northeastern China

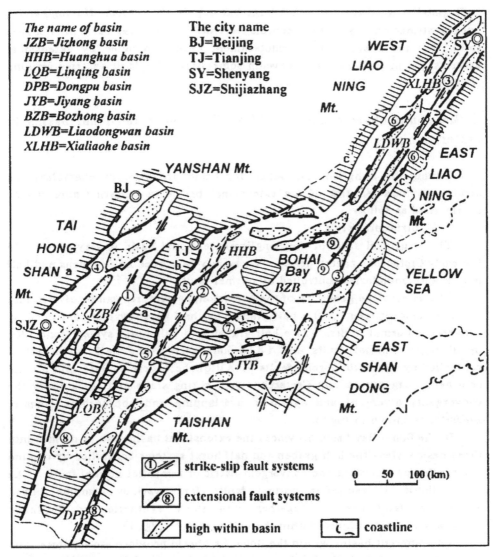

The name of basin
JZB=Jizhong basin
HHB=Huanghua basin
LQB=Linqing basin
DPB=Dongpu basin
JYB=Jiyang basin
BZB=Bozhong basin
LDWB=Liaodongwan basin
XLHB=Xialiaohe basin

The city name
BJ=Beijing
TJ=Tianjing
SY=Shenyang
SJZ=Shijiazhang

strike-slip fault systems
extensional fault systems
high within basin
coastline

0   50   100 (km)

The serial number of the major fault systems: ①-the right-lateral strike-slip fault zone from Baxian to Shulu to Handan; ②-the right-lateral strike-slip fault zone from Huanghua to Dezhou to Dongming; ③-the right-lateral strike-slip fault zone from Shenyang to Weifang; ④-the extensional fault system in the Jizhong rifting basin; ⑤-the extensional fault system in the Huang hua rifting basin; ⑥-the extensional fault system in the Xialiaohe-Liaodongwan rifting basin; ⑦-the extensional fault system in the Jiyang rifting basin; ①-the extensional fault system in the Linqing-Dongpu rifting basin; ⑨-the extensional fault system in the Bozhong rifting basin

Fig.2   Sketch map of the Cenozoic tectonics in Bohai Bay basin province

province resembles a large pull-apart basin. In fact, most of the Paleogene faulted basins are controlled by normal faults striking NE-NNE, so that many geologists[1] regard the basin as a large Palaeogene rift system. Having carried out a systematic study on the Cenozoic tectonics in Bohai Bay basin provice, we think that the basin is a supradetachment basin controlled by extensional systems and in the basin also develop some strike-slip fault systems striking NNE[2].

2. Structural charcteristics of Meso-Cenozoic extensional basins in eastern China

Taking the Songliao and Bohai Bay basin provinces as examples, the structural characteristics of the extensional basins in eastern China can be summerized as follows:

2.1 Structural style

The structural style of the Songliao extensional basin is very typical. In the early Cretaceous rifting period were mainly developed half grabens and half horsts with boundaries controlled by normal faults.

The assemblage geometry of the extensional structures can be divided into three tectonic systems: (1) a half graben and half horst system controlled by convergent normal faults; (2) a half graben and half horst system controlled by divergent normal faults and (3) a half graben and half horst system controlled by synthetic normal faults[3]. All of them were evolved into a unite continental sag during late Cretaceous post-rifting stage and adjusted by the convergent, divergent and synthetic overlapping and approaching transfer zones[4], as shown in Fig. 3.

In the Bohai Bay basin province, the extensional basins can be divided into three basic styles: the half graben and half horst systems controlled by domino (or synthetic) divergent and convergent listric or ramp-flat normal faults. The major listric or ramp-flat normal faults form various complex linked extensional fault systems together with those synthetic and antithetic secondary faults developed within the half graben[5](Fig. 4).

The different basins within the Bohai Bay basin province are separated and linked by some regional transfer zones which include NW and EW transfer faults and transverse highs. The extensional tectonics takes a extensional direction of NWW-SEE and penetrates through the upper crust of the basin province and detaches within the middle crust about a depth of 15—20km.

2.2 Tectonic subsidence

The subsidence curve of the extensional basins consists of two sections which are clearly different. Take the Songliao basin as an example, the first section is very steep, representing K_1 rifting stage, which indicates that the rate of the total basement subsidence and   tectonic subsidence is relatively high . On the other hand, the second one shows a gentle curve of the total basement subsidence and tectonic subsidence, and represents the cooling period

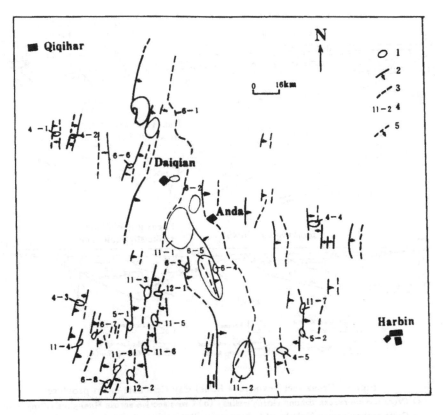

Fig. 3   The schematic map of transfer zones in the northern Songliou basin
1. Transfer zones; 2. Normal faults; 3. Gentle slop;
4. Number of transfer zones; 5. Inferred normal faults.

in late Cretaceous, showing that the rate is relatively low. The two sections constitute an upward concave curve (Fig. 5).

2. 3   Modification of the extensional basins

The extensional basins have been modified by horizontal compression and strike-slip faulting. For example, the Songliao basin province contains half graben systems in the lower part and large fault propagation anticlines in the upper part. The Lindian structure belongs to this type [3] (Fig. 6).

In Bohai Bay basin province, the Tertiary extensional basins were modified by NNE and NE deep strike-slip faulting, which have evidently been moved right-laterally during late Cenozoic. In response to the movement pattern, the cover of basin overlying the deep faults had a right-lateral strike slip deformation. There are three regional NNE right-lateral strike-slip faults superposed upon the extensional tectonics which consists of the NE-NNE normal faults and the NW-NWW-EW transfer faults. The names of the strike-slip zones from west to east are Baxian-Shulu-Handan, Huanghua-Dezhou-Dongming and Shenyang-Weifang respectively (Fig. 2, Fig. 4). Among these strike-slip zones, Shenyang-Weifang strike-slip zone is the clearest one. Some

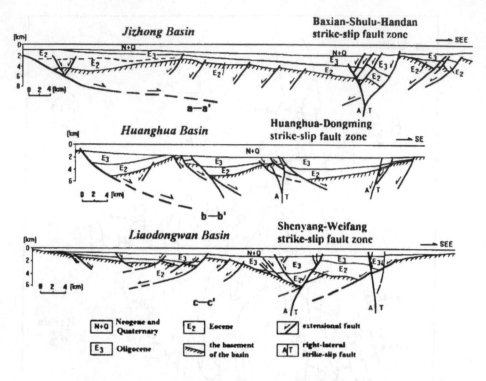

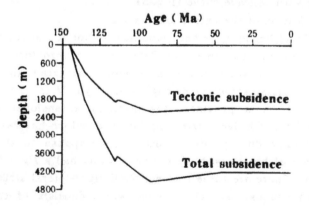

Fig. 4 Cross sections across Bohai Bay Cenozoic basin province
(a) Cross section across Jizhong rifting basins; (b) Cross section across Huanghua rifting basin;
(c) Cross section across Xialiaheo-Liaodongwang rifting basin. 1—Neogene and Quaternary; 2
—Oligocene (Dongying Formation and the 1st—2nd member of Sahejie Formation); 3—Eocene
(the 3rd—4th member of Sahejie Formation and Kongdian Formation); 4—the basemant of the
basins; 5 — normal faults; 6 — strike-slip fault (A: displacement away from viewer, T:
displacement toward viewer). The place of the section see Fig. 2.

Fig. 5 Tectonic subsidence and total subsidence curves of basement
in the Lishu fault depression of southeasten Songliao basin.

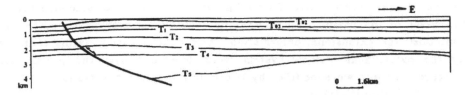

Fig. 6   Seismic line from the Lindian faulted depression of the Songliao basin,
showing a half graben in the lower part and inversion structure in the upper part.

T_5—T_4: Shahezi and Yingcheng Formations of K_1; T_4—T_3: Denglouku Formation.
T_3—T_2: Quantou Formation of K_1; T_2—T_1: Qingshankou Formation of K_2; T_1—
T_{03}: Sifangtai Formation of K_2; T_{03}—T_{02}: Mingshui Formation of K_2.

peculiar strike-slip faults and associated structure elements are distributed
along the east edge of Liaodongwan basin (fig. 4c, fig. 7) and Bozhong basins
(Qi. et al. , 1992)[2].

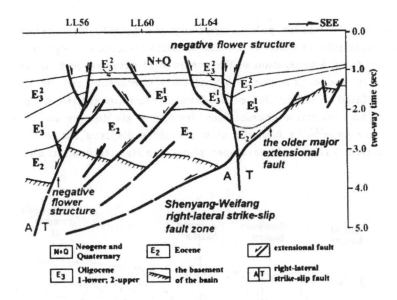

Fig. 7   Cross section across the east edge of Liaodongwan basin
1—Neogene and Quaternry; 2—Upper Oligocene (Dongying Formation); 3—Lower Oligocene
(the 1st — 2nd member of Sahejie Formation); 4—Eocene (the 3rd — 4th member of Sahejie
Formation and Kongdian Formation); 5—The basment of the basins; 6—normal fault; 7—
strike-slip fault (A: displacement away from viewer. T: displacement toward viewer).

On the section across the east edge of the Liaodongwan basin, the steep
strike slip faults dissect the low angle extensional faults and display the style
of a negative flower structure.

Not as the extensional tectonics which disperse the whole basin province,
the NNE strike-slip tectonics are a zonal or linear tectonic system which is

present within a narrow belt of 1—10km (including the associated structures in the cover of the basin). The strike-slip tectonic systems also modified and controlled the extensional deformation. The Eogene rifting basins controlled by the extensional tectonic system were developed over the pre-Cenozoic fracture belts and were modified by late-Cenozoic strike-slip faults.

2.4 Volcanic activity

Volcanic rocks were developed in late Jurassic pre-rifing and early Cretaceous rifting stages, and in rifting and post-rifting stages of Tertiary extensional basins.

The petrological and geochemical characteristics of late Jurassic and early Cretaceous volcanic rocks are different from those of Tertiary. The former is composed of high-K calc-alkaline and shoshonite series. whereas the latter consists of alkaline and tholeiitic series.

3. Dynamics of the Meso-Cenozoic extensional basins

Taking all of these point, especially the petrological characteristics of volcanic rocks into account, and based on plate tectonic settings, two modes of mechanism for the generation of the continental rifts can be proposed.

(1)First mode. The formation of early Cretaceous rifts in northeastern China is related to subduction of the Isunazi plate toward the Asian continent (Fig. 8). The subduction characterized by low angle dipping slab is similar to the Andeans type[6]. In fact, the slab geometry displays listric form and the early Cretaceous rifts were developed above the flat segment, as shown in Fig. 8.

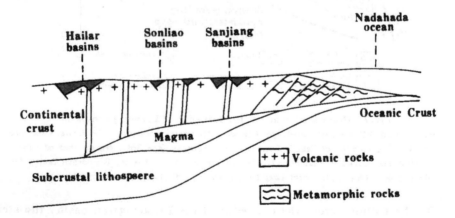

Fig. 8 Geodynamic model of late-Mesozoic continental rifts in northeastern China.

As evidenced by the presence of high-K calc-alkaline and shoshonite series of volcanic rocks related to subduction, thick continental crust formed during pre-rifting shortening and subsequent regional extension, it appears that these early Cretaceous rifts were resulted from extensional collapse of orogenic zones[7] and enhanced by roll back of subduction boundary [8]

(2) Second mode. The formation of Tertiary basins in eastern China is related to subduction of the Kula and Western Pacific plates. The subduction is typified by steep angle dipping slab and is similar to Mariana type. As evidenced by the presence of alkaline series of volcanic rocks, significant regional extension and roll back of trenches, these Tertiary extension basins, such as East Sea basin, northern Jiansu basin, Bohai Bay basin and others are thought to evolve in response to decreasing convergence rates or divergence of the colliding plates.

As regards the role of mantle plume in basin formation, we suggest that it is secondary and is induced by regional extension.

How does the Tertiary extensional tectonic systems make the coordinated movements with the striks-slip tectonic systems? we propose an explanatory model for the kinematic mechanism of the Bohai Bay basin province. In the model (Fig. 9), the extensional systems in the basin are thin-skinned tectonics. The regional detachment faults divorce the upper crust from the lower crust and lithospherical mantle, so in the supradetachment crust can be produced the extensional displacement by block-faulting, and in the underdetachment crust and mantle produced the extensional displacement by ductile stretching, whereas in the place of the strike-slip zones, the strike slip systems keep close relations with the deep faults and become a kind of vertical tectonic zones through the whole lithosphere.

The extension tectonices and the strike-slip tectonics may be resulted from

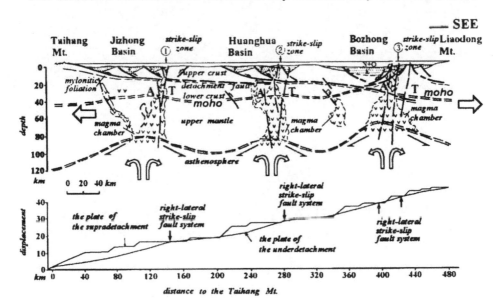

Fig. 9   An explanatory model for the kinematic mechanism of the Bohai bay basin province

the different dynamics. The driving force of NE or NNE extensional tectonics may be originated from horizontal tension of NW-SE direction during Eocene and Oligocene, whereas the NNE strike-slip tectonics may be resulted from the continent transform movement induced by horizontal tension of N—S direction along NNE pre-existing Tan-Lu faults during late Oligocere, Neogene and Quarternary.

References

1. Ma Xinyuan, Liu Hepu, Wang Weixiang et al. , Meso-Cenozoic taphrogenesis and extensional tectonics in the East China. Acta Geologica Sinica, 1983, Vol. 57, No. 1 (In Chinese with English abstract).
2. Qi Jiafu and Chen Fajing, Structural style in Liaodongwan-Xialiaohe basin. Oil and Gas Geology, 1992, Vol. 13, No. 3, pp. 272—283 (In Chinese with English abstract).
3. Chen Fajing, Wan Xinwen, Zhang Guangya et al. , Tectonic and dynamic scenario of Meso-Cenozoic petroliferous basins in China, Modern Geology, 1992, Vol. 6, No. 3, pp. 317—327 (In Chinese with English abstract).
4. Morley C. K. , Nelson R. A. , Patton T. L. , et al. , Transfer zones in the East African Rift system and their relevance to hydrocarbon exploration in rifts. A. A. P. G. , 1990, Vol. 74, No. 8, pp. 1234—1253.
5. Gibbs, A. D. , Linked fault families in basin formation, J. Struct. Geol. , 1990, 4, pp. 795—803.
6. Mechel sebrier, Jacques Lowis Mercier, Jose Machare et al. , The state of stress in an overriding plate situated above a flat slab; The Andes of central Peru, Tectonics, 1988, Vol. 7, No. 4, pp. 895—928.
7. Dewey J. F. , Extensional collapse of orogens, Tectonics, 1988, Vol. 7, No. 6, pp. 1123—1139.
8. Leigh H. Royden, The tectonic expression slab pull at continent convergent boundaries, Tectonics, 1993, Vol. 12, No. 2, pp. 303—325.

Proc. 30ᵗʰ Int'l. Geol. Congr., Vol. 6, pp. 47-53
Xiao Xuchang and Liu Hefu (Eds)
© VSP 1997

The Tectonic Evolution, Basin Development and Rock-stratigraphic Sequence of Jurassic-Cretaceous Period in North China

Yigang Bao,   Zhenfeng Liu

Beijing Geological Survey Institute, China

The distribution of Jurassic-Cretaceous strata in North China is showed as Fig.1. According to their occurrence and characteristics, the strata are divided into 7 subdivisions, i.e. Ordos, Shanxi, Yanliao-Yinshan, the west of Shandong, Henan and Anhui, the east of Liaoning and the east of Shandong.

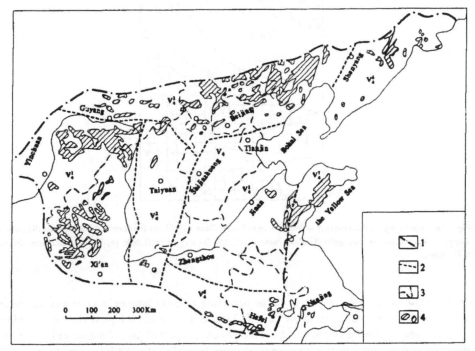

Fig.1 Sketch map of crop distribution and stratigraphic subdivision of Jurassic-Cretaceous in North China. 1. North China Platform boundary; 2.The stratigraphic subdivision boundary; 3.The provincial boundary; 4. The crop regions of Jurassic-Cretaceous strata; V_4: North China stratigraphic division; V_4^1: Ordos subdivision; V_4^2: Shanxi subdivision; V_4^3 :Yanliao-Yinshan subdivision; V_4^4 : Subdivision of the west of Shandong; V_4^5 : Subdivision of Henan and Anhui; V_4^6 : Subdivision of the east of Liaoning; V_4^7 : Subdivision of the east of Shandong.

As to the general tectonic background of this region during Jurassic-Cretaceous, Chinese geologists have gained the following common knowledges:

(1) Indo-sinian Movement occurred from Late Permian to Late Triassic. Its concrete behaviour was that Siberia Platform moved southwards and produced compressive stress field. Therefore around

the south margin of Siberia shield-shaped Oldland formed an arc tectonic system (Fig.2). For example, Mongolia-Xing'an Indo-sinian folding belt and Qinling Indo-sinian orogenic belt.

In the interior of North China Platform, Indo-sinian Movement did not act intensely except within some areas. Ordos and Shanxi areas merely uplifted as a whole and had a disconformity in it.

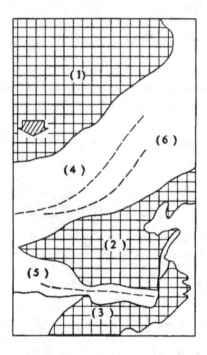

Fig.2 Tectonic background sketch map of Indo-sinian in North China. (1) Siberia Platform; (2) North China Platform; (3) Yangtze Paraplatform; (4) Mongolia-Xing'an orogenic belt; (5) Qinling orogenic belt; (6) The structural line strike of Indo-sinian.

(2) From the end of Late Triassic or in the beginning of Early Jurassic, a great tectonic system transformation appeared in the eastern part of China, and the main geological event is the oblique collision between the West Pacific Oldland and the eastern part of Asian Continent(Fig.3) (quoted from Jishun Ren, 1990). Such a dynamic mechanism formed a set of NNE trending fault series in the eastern part of China and resulted in a series of terrestrial basins and tectonic-magmatic mobile belts in a large scale. The peak of this tectonic activity was from Late Jurassic to the beginning of Early Cretaceous. This great transformation and the various important geological events derived from it are called Yanshan Movement in China.

On the basis of the achievements of regional research in the past few years, three important viewpoints are put forward in this paper.

1. Basins and their generations

The terrestrial basins of Jurassic-Cretaceous in North China Platform were restricted by the different tectonic evelution stages and can be divided into 4 basining stages. Different generation basins were developed in a certain tectonic positions and have their own rock associations(Fig.4).

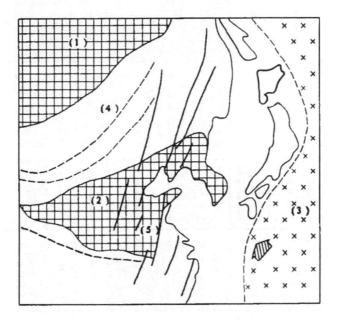

Fig.3 Tectonic background sketch map of Yanshanian in North China (by Jishun Ren, 1990). (1) Siberia Platform; (2) North China Platform;(3)The West Pacific Oldland which had split and sunk; (4) The structural line strike of Indosinian ; (5) The giant shear fracture series of Yanshanian.

The first generation basins were in Early-Middle Jurassic. After the stress relaxation of Indosinian Movement in Late Triassic, formed the inherited down-warped basins.

Ordos and Shanxi areas were stable sedimentary regions from Late Paleozoic to Early Mesozoic, so they inheritedly developed into great stable introcontinental down-warped basins at this stage. However, the basins in the eastern part of North China Platform were small and unstable. They were mainly developed in the northern and southern sides of the Platform. There were no first generation basins in the most eastern part of the Platform,i.e.the subdivision of the east of Shandong.

The strata consisted of three parts. The lower part is macroclastic rock,there were intermediate-basic volcanic eruptions in Yanliao area at the same time, forming volcanic rock units named Xinglonggou Fm or Nandaling Fm.The middle is a set of coal-bearing microclastic rocks. The upper is mottled clastic rocks.

The tectonic movement of the first phase of Yanshan Movement made the first generation basins to die out.

The second generation basins were developed in Late Jurassic. Being in the radical period of tectonic system transformation, North China Platform was uplifted as a whole, only in the northern, western and southern sides of the Platform were developed the second generation graben basins.

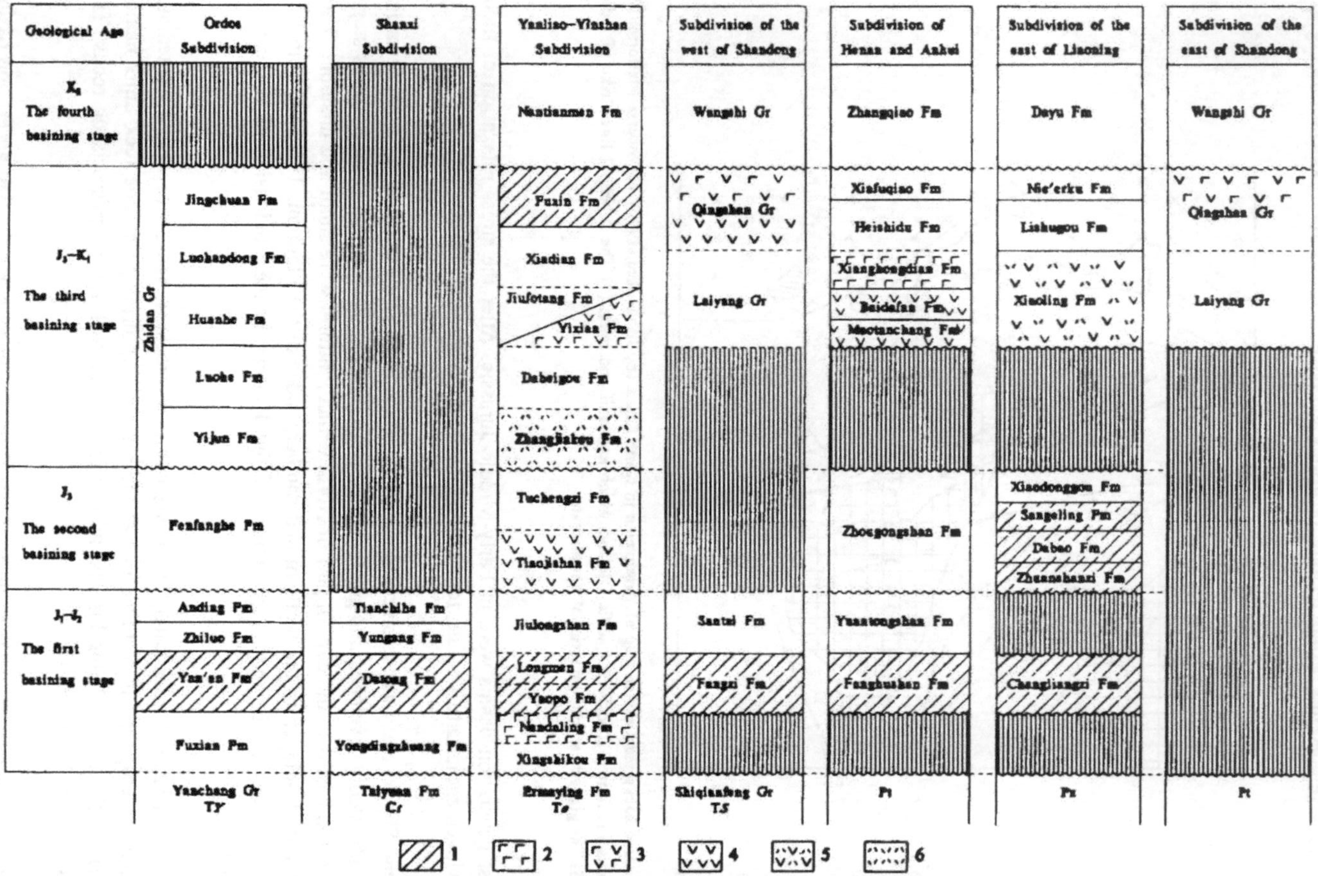

Fig.4 Diagram showing the lithostratigraphic sequences and the subdivision of basining stages of Jurassic—Cretaceous Period in North China. 1.coil-bearing; 2.basalt; 3.basalt–andesite; 4.latite; 5.rhyolite–andesite; 6.rhyolite

Their strata are mainly fast accumulated macroclastic rocks. Lithology and thickness change sharply in horizontal directions. The main tone is brownish red, greyish purple, yellowish brown, which indicating the intense oxidative environment. The volcanic activities were only developed in Yanliao area, the rocks are trachyandesite and andesite.

Compared with the first generation basins, the greatest difference between them is that the dark colour coal-bearing microclastic rocks were not developed in the middle part of the basin accumulation sequence in all subdivisions except the eastern part of Liaoning. The second generation basins did not last long and ended in the latest Late Jurassic.

The great tectonic system transformation was accomplished in the latest Late Jurassic, a new tectonic pattern was formed.The giant NNE trending shear fracture series in East China played a leading role, controlled and resulted in the third generation basins, the volcanic eruption belts and intrusive rock belts. Their directions are NNE trending as well.

Most of the third generation basins were neogenic,the strata usually overlay unconformably on the basements of different ages. The volcanic activities were intense and widespread, especially in the eastern part of Zijingguan Deep Fracture.

In the later of this basining stage occurred again a set of coal-bearing strata called Fuxin Fm(or Qingshila Fm, Guyang Fm). They distributed roughly in the north of present 41 ° north latitude. In the south of this latitude, the basins were dominated by an arid environment, forming clastic rock series with intense oxidative tone.

After the third generation basins ended at the end of Early Cretaceous,the Platform underwent a new great tectonic system transformation, the direction of maximum principal compressive stress turned to NE trending, therefore the pre-existed NNE trending sinistral compresso-shear faults turned to tensile dextral strike-slip which in turn resulted in the development of the fourth generation graben basins.

The basins were mainly developed in the eastern part of North China Platform. The rocks are distinctive brick-red clastic sediments, they are not compact but relative loose.

The basining stage began from Late Cretaceous and could last to Paleogene. It already belonged to Himalayan Cycle.

2. The characteristics of volcanic rocks

There were three peaks of volcanic activities.They occurred respectively in the early-middle stages of the first, second and third basining stages.

The first and second activities distributed rather limitedly,only occurred in a few volcano-sedimentary basins of Yanliao area. The volcanic activities within the third basining stage not only reached the maximum value of intensity, but also distributed in the eastern part of the Platform which was equal to the broad areas of the eastern part of Zijingguan Deep Fracture. The volcanic rock belts had strong orientation. They were NNE trending in general, and were closely related to the NNE trending giant shear fault series.

By thorough and systematic research on the volcanic geology, volcanic rock and volcanic process in Beijing and the northern part of Hebei Province recently, a new knowledge about the volcanic rock type and volcanic rock series type is gained.

It was generally believed in the past that the volcanic rock type in this region was mainly andesite and basalt-andesite. According to the total alkali sillica classification, however, the volcanic rocks consist of basaltic rock, shoshonite, latite, trachyte and high-potassium rhyolite, among which the latite is dominant. The authors think that the latitic series is an independent volcanic rock series, which may coodinate with the tholeiite series,calc-alkaline series and alkali-basalt series. Seen from Fig.5, the MORB normalized distribution patterns are most similar to that of WPB.It can be proved that the volcanic rocks had formed in the interior of continental plate, the volcanic activities were controlled by introplate fault orogeny.

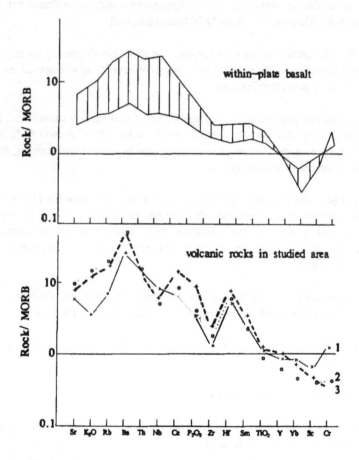

Fig.5 MORB normalized distribution patterns of basalts(by Pearce,1982). 1. Nandaling basalt; 2. Tiaojishan basaltandesite; 3. Donglanggou shoshonite.

3. Southward slipping of North China Platform during Yanshan tectonic cycle

Most of the modern coal-bearing strata distribute in the high latitude areas of more than 45 °

north latitude. They belong to the temperate zone and are good marks to indicate the climate. Of the coal-bearing strata occurred in the first generation basins in North China Platform(Fig. 6), the most southward ones distributed along the line of present 32 ° -33 ° north latitude. The second coal-bearing basins of Late Cretaceous occurred in the north of present 41 ° north latitude. The difference of the two lines is up to 9 ° latitude. It shows that during the oblique collision between the eastern part of Asian Continent and the West Pacific Oldland, North China Platform, as a part of Asian Continent, slipped southwards for 9 ° latitude with the horizontal distance no less than 1 000 km in Yanshan tectonic cycle from Early Jurassic to the end of Early Cretaceous.

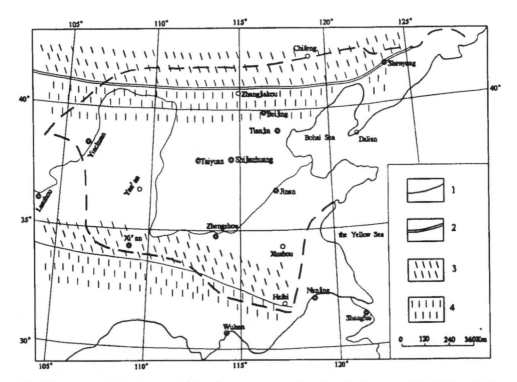

Fig.6 The map showing the changes of palaeoclimate subzone location of Jurassic-Cretaceous in North China. 1. The palaeoclimate boundary in Early-Middle Jurassic; 2. The palaeoclimate boundary in Early Cretaceous; 3. The temperate damp climate zone; 4. Tropical-subtropical arid-semiarid climate zone.

REFERENCES

1. Zhimin Bai, Yigang Bao and Shiwei Ge. Latitic series of the Yanshanian tectonic cycle in Beijing region. *Journal of Geology and Mineral Resources of North China*. Vol. 10, No. 1, 51 ~ 63.(1995)
2. Jishun Ren et al. Tectonic evolution of the continental lithosphere and metallogeny in eastern China and adjacent areas. Beijing: Sciences Press.(1990)

Proc. 30ᵗʰ Int'l. Geol. Congr., Vol. 6, pp. 55-76
Xiao Xuchang and Liu Hefu (Eds)
© VSP 1997

Tethyan Evolution in South China and Its Environs

WU GENYAO

Laboratory of Lithosphere Tectonic Evolution, Institute of Geology, Chinese Academy of Sciences, Beijing 100029, China

Abstract

Asian continent is a composite one, resulted from amalgamation of a number of microcontinents and blocks which were once separated by oceans and troughs. The microcontinents can be grouped as four regiments: the Yangtze affinity, the Australian Gondwana affinity, the Australian Gondwana and the Indian Gondwana. Three Paleotethyan geosuture zones developed in southwest China: the Bitu-Changnjng-Menglian, the Jinshajiang-Mojiang and the Ganzi-Litang, with the former representing the subducted major oceanic basins. The closures of Paleotethys and followed collisions create an integrated land called the China-SE Asian Subcontinent. Two Neotethyan geosuture zones occurred in Xizang (Tibet): the Bangonghu-Dengqen, and the Indus-Yarlung Zangbo. The sinistral Changle-Nan'ao fracture belt in the southeast coast region of China is an oblique collision and shear orogeny between South China and the Donghai Microcontinent. The final closure of Neotethyan oceans, the collison and weld of the Indian Craton, China-SE Asian Subcontinent and the Siberian Craton bring Asia continent into being.

Keywords: Tethyside orogenic collage, Neotethys, Paleotethys, Prototethys, South China and its environs

INTRODUCTION

The well-know term " Tethys " was coined by E. Suess in his monumental work " The Face of the Earth". For a long time it was thought to be a sea, similar to the present Mediterranean Sea, which existed along the general course of the Alpine - Himalayan orogenic belt: a composite geosyncline out of which many of the structures of the present orogenic belt were formed; and the composite geosyncline occurred from the Permian into the Early Tertiary (cf. Huang Jiqing and Chen Bingwei[31]).

South China is a key area for the study of Himalayan orogenic belt. The ophiolites discovered in the Qinghai - Xizang (Tibet) Plateau and the Sanjiang (Three Rivers, namely, the Nujiang-Salween, the Lancang and the Jinsha Rivers) Area in 70's and early 80's[18, 24, 29, 47, 53, 54, 67, 69, 88, 93, 97, 100] caused a new idea that Tethys must have been an ocean between the Gondwana and Eurasia continents. The faunas as well as sediments of Gondwana facies determined during that time[19, 37, 41, 44, 70, 89, and for more references, see 6, 64, 71, 91] provided further evidences of Tethys being an important demarcation for paleobiogeographycal provinces.

In more recent years, the research on Tethys in South China, especially in the Sanjiang Area which was once regarded as the Indosinian geosyncline, has made significant contributions[36, 45, 48, 52, 82, 94, 95, 99, 103]. Zhong Dalai et al.[105] put forward that the eastern Tethys might be a wide and complicated marine system which contained intercontinental oceans, intermicrocontinental oceans and intracontinental troughs or aulacogens. The Tethyan evoluiton could be divided into three major cycles, namely, Prototethys from Sinian to Silurian, Paleotethys from Devonian to Triassic and Neotethys from Jurassic (or the latest Triassic) to Tertiary[80, 104]. It is these works that make it possible to reunderstand the tectonic framework of South China and its environs and to reconstruct the tectonic history related to Tethyan evolution.

GEOSUTURE ZONES AND MICROCONTINENTS

Paleotethyan Geosuture Zones

Three Paleotethyan geosuture zones can be distinguished in the Sanjiang Area, which are called, in an eastward order, the Bitu-Changning-Menglian, the Jinshajiang-Mojiang and the Ganzi-Litang geosuture zones, with the former representing the subducted major oceanic basins (Figs.1, 2). The Bitu ophiolite distributes along the Yuqu River, east Xizang[56, 81]. Down to the northwest Yunnan, the dismembered ophiolite suites are discovered in the Lancang River valley[49, 52]. The oceanic records in southwest Yunnan are represented by the Tongchangjie ophiolitic melange in Yunxian[93], the pillow lava (MORB) and chert to the south of Menglian[15, 73]. In addition, the glaucophane schist which marks a high pressure metamorphism related to oceanic subduction is well exposed in the Changning-Menglian belt[57, 98, 102, 107]. The ophiolitic melange of Jinshajiang Oceanic Basin is best exposed in the southern section bordering Sichuan, Yunnan and Xizang, along the Jinsha River valley. The recent work by Sun Xiaomeng et al.[66], based on geochemical characteristics of radiolarian chert, suggested that the bloom period for oceanic crust development was in Early Permian. The ultramafic rocks, basalt and chert in the north sector of the Ailao Mt., an example from the Shuanggou ophiolite in Xinping[96], are geological records of the Mojiang Oceanic Basin. The dismembered ophiolite suites of the Ganzi-Litang Oceanic Basin stretches from Yushu, Qinghai Province, southwards through Ganzi, Litang and Muli, Sichuan Province, to Luoji, Zhongdian, Yunnan Province[95]. The glaucophane schist is also discovered in the ophiolitic melange belt[63].

To the west of the Bitu-Changning-Menglian zone is the Shan-Thai (Sibumasu) Microcontinent, which might be an outlying region of Gondwana during Late Paleozoic. To the east of the Ganzi-Litang zone is the Yangtze Microcontinent, which was cratonized by the end of Archean, and enlarged by collision and accretion of the Xianggan and Cathaysia Microcontinents in Late Proterozoic and Early Paleozoic[59]. Between the Bitu, Jinshajiang and Ganzi-Litang oceans were the Chaya-Mangkang and Yidun Blocks respectively. The both should be once a part of the Yangtze Microcontinent, since the Mid Proterozoic, Sinian and Lower Cambrian of the Yangtze-type were discovered in the Jinsha River valley bordering Sichuan and Xizang[21, 50]. Between the Changning-Menglian and Mojiang oceans existed the Simao Block, which might be northward extension of the Indosinian Microcontinent, and dispersed from the latter by the opening of the Nan River Oceanic Basin[51].

Stretching northwestwards, all of the three above-mentioned geosuture zones appear to be merged into the Longmucuo-Yushu fracture. There are different opinions on the basalt and mafic-ultramafic rocks outcropping along the fracture. Some persons[9, 38, 40] regarded them ophiolite suite and the fracture a geosuture zone representing the north boundary of Gondwana; others thought them being of intra-plate basalt[28]. In this paper the fracture is tentatively regarded as a plate boundary. As to

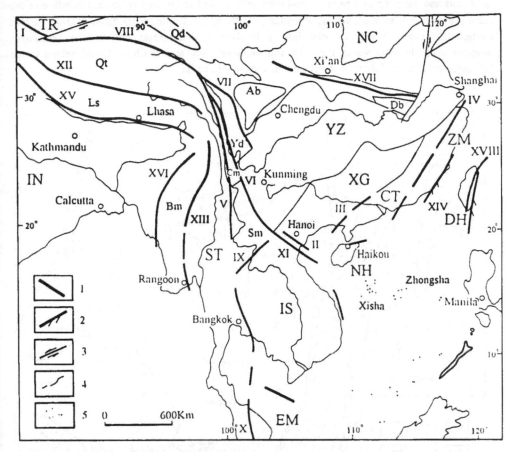

Figure 1. Tectonic framework of South China and its environs

1. Phanerozoic geosuture zone (or plate boundary); 2. Oblique convergent and collisional zone; 3. Shear zone; 4. Boundary of microcontinent or block; 5. Islands; Name of the microcontinents: CT. Cathaysia; DH. Donghai; EM. East Malaya; IN. India plate; IS. Indosinia; NC. North China; NH. Nanhai; ST. Shan-Thai; TR. Tarim; XG. Xianggan; YZ. Yangtze; ZM. Zheming; Name of the blocks: Ab. Aba; Bm. Burma; Cm. Chaya-Mangkang; Db. Dabei; Ls. Lhasa; Qd. Qaidam; Qt. Qiangtang; Sm. Simao; Yd. Yidun; Name of the geosuture zones or plate boundary fractures: I. Western Kunlun Mt.; II. Song Da; III. Yunkai; IV. Lishui-Zhenghe-Dapu; V. Bitu-Changning-Menglian; VI. Jinshajiang-Mojiang; VII. Ganzi-Litang; VIII. Longmucuo-Yushu; IX. Nan River; X. Bentong-Raub; XI. Song Ma; XII. Bangonghu-Dengqen; XIII. Mandalay-Jade Mines; XIV. Changle-Nan'ao; XV. Indus-Yarlung Zangbo; XVI. Naga-Manipur; XVII. Qingling; XVIII. Central Range, Taiwan

southward elongating of the geosuture zones is concerned, the Bitu- Changning-Menglian zone can be well connected to the Bentong-Raub geosuture zone in central Malaysia[35], which separates the Shan-Thai from the East Malaya Microcontinents. The southeast extension of the Jinshajiang-Mojiang zone is the Song Ma geosuture zone in north Vietnam[35], which is a boundary between the Indosinian and the Yangtze - Xianggan Microcontinents. Eastwards further, the Triassic abyssal deposits and Late Paleozoic basites, well developed in west Guangxi, were regarded as Paleotethyan records by Wu Haoruo et al.[86]. Furthermore, the bathyal- abyssal Carbono-Permian sediments and Early Triassic turbidite and contourite as well as bimodal submarine eruption in the coast region of Fujian and Zhejiang Provinces[22, 26, 43], indicating a southeast-dipping passive continental margin, which inverted into a foreland fold-and-thrust belt in the Indosinian orogenesis[27]. In other words, the south and southeast coast area of China should belong to the Paleotethyan domain, only the ophiolite not discovered yet so far.

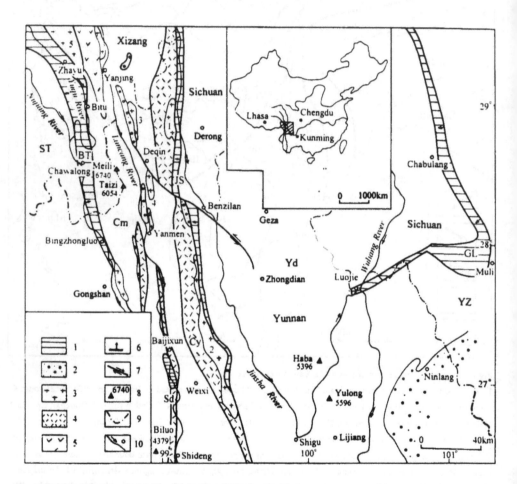

Figure 2. Main tectonic units in the central sector of the Sanjiang Indosinides bordering Xizang, Sichuan and Yunnan, the inset showing location of the region investigated (shadowed)

1. Ophiolitic melange; 2. Molasse; 3. Collision-type granitoid; 4. Volcanic rocks in delamination-related rift; 5. Middle Triassic volcanics (the Zhuka formation); 6. Subduction vergence; 7. Strike-slipping fracture; 8. Peak and elevation; 9. Provincial boundary; 10. River and town; Name of the tectonic units: BT. The Bitu geosuture zone; Cm. The Chaya-Mangkang Block; GL. The Ganzi-Litang geosuture zone; JS. The Jinshajiang geosuture zone; ST. The Shan-Thai Microcontinent; Yd. The Yidun Block; YZ. The Yangtze Microcontinent; Cy. The Cuiyibi Rift (east branch of the Northwestern Yunnan Rift System) ; Sd. The Shideng Rift (west branch of the Northwestern Yunnan Rift System); Name of the granitoids: 1. Jiaren; 2. Ludian; 3. Adengge; 4. Baimang; 5. Dongdashan

Neotethyan Geosuture Zones

There are two Neotethyan geosuture zones in the Qinghai-Xizang Plateau and its adjacent regions (Fig.1). One is named the Bangonghu-Dengqen geosuture zone, which separated the Qiangtang Block from the Lhasa Block. Both of the blocks made up the North Xizang Microcontinent, which should be originated from the Indian Gondwana. Its southward extention may be the belt of Mandalay-Jade Mines in Central Myanmar[35] which, marked with ophiolite and high pressure metamorphic rocks, represented a disappeared ocean between the Burma Block and the Shan-Thai Microcontinent in Mesozoic, and acted as a dextral strike-slipping fracture since Late Tertiary, namely, the Sagaing fracture[23]. Another is the Indus-Yarlung Zangbo-Naga-Manipur geosuture zone which constitutes the north boundary of the India plate.

The sinistral Changle-Nan'ao fracture in the coast region of Fujian may represent an oblique convergence and collision between the amalgamated Huanan (South China) Subcontinent and the Donghai Microcontinent. The unmetamorphosed Lower Cretaceous depositing on the strata in both sides of the fracture implies the oblique collision and shear orogeny ceased in the earliest Cretaceous.

SOME CLUES FOR PROTOTETHYS

The western Kunlun Mt. belt in northwest part of the Qinghai-Xizang Plateau is identified as a geosuture zone of Prototethys. The oceanic basin, which is marked by the Kudi-Shubashi ophiolite in southern Xinjiang[12, 13], is closed by the end of Ordovician and subducted southwards[55]. The isotopic datings (480-384 Ma) and geochemical features provide further evidences for the ocean and indicate the geosuture zone can extend for a thousand of kilometers[87]. Pan Yusheng[55] regarded it as a Prototethyan track since the ocean is located between the Angara and Gondwana and marked a southern boundary of the Paleo-Asian continent during Sinian-Ordovician. In the author's opinion, the major oceanic basin of Prototethys might be situated in the Tianshan-South Mongolia area.

Whether the Prototethys occurred in the Sanjiang Area is still to be clarified by further work. However, the following observations support the thought that the Bitu - Changning - Menglian Oceanic Basin may inherit an Early Paleozoic abyssal trough or oceanic basin. 1) The Ordovician-Silurian continental slope deposits are well developed along both of the east margin of the Shan-Thai and the west one of the Yangtze Microcontinents. 2) The Lower Devonian sediments of deep marine facies are continuously deposited on the Silurian graptolite shale[74]. 3) The Early Paleozoic faunas (trilobite, Cephalopoda, etc.) in the Shan-Thai domain are different from those in the Yangtze area[103]. 4) The blue schist in Lancang dated about 410 Ma[11] might be a record of high pressure metamorphism related to the subduction of oceanic crust. 5) The calc-alkaline rocks of volcanic arc sandwiched in the Lower Paleozoic and subduction-type tonalite in an age of 462 Ma occurred in Jiangda, northeast Xizang[39], and the andesitic lithoclast discovered in the Devonian deposits in Simao, southwest Yunnan, comprehend an active continetal margin. In addition, the subduction-related adamellite in an age of 456 Ma is reported by Zhang Yuquan et al.[99]. 6) An unconformity between the Devonian and the underlying beds[7] implies an orogenesis occurring before Devonian, although no ophiolite being found so far.

The Song Da belt in north Vietnam representing a Prototethyan geosuture zone[35] has been widely accepted. Whether or not it may stretch eastwards to the middle-south area of China has been argued. Sun Mingzhi and Xu Keqin[65] put forward that an active continental margin, with assemblage of tonalite-plagiogranite-granodiarite and in an age of 552-410 Ma, was existed along the Yunkai-Wuyi belt during Early Paleozoic. Qiu Yuanxi[58] reported the arc-related basalt and granitoids in an age of Sinian-Early Paleozoic in the Yunkai Mt. region, west Guangdong. To echo each other over a distance, the Xiongshang diabase swarm, with petrochemistry characters of MORB, tectonic setting of back-arc basin and in an age about 590 Ma, was discovered by Ren Shengli et al.[61] at Zhenghe County, north Fujian. The Lishui-Zhenghe-Dapu fracture, to the west of which is the Cathaysia Microcontinent and to the east the Zheming (abbreviated from Zhejiang and Fujian Provinces) Microcontinent, must have been a plate boundary. It is thought, tentatively, to be a Paleotethyan geosuture zone in this paper, for few Late Sinian-Ordovician strata and Early Paleozoic granitoids occur in the Zheming Microcontinent, which is obviously different from those in Cathaysia. Furthermore, the fusulinid, coral, brachiopod and bryozoan faunas of Gondwana affinity, for example, *Monodiexodina*, are discovered in the Zheming Microcontinent[5], which means the microcontinent should be of Australian Gondwana affinity during Late Paleozoic. Another possibility for the fracture is that it is a geosuture zone of Prototethys, but the block, previously to the east of the fracture, has underthrusted or distorted beyond recognition by younger tectono-thermal events.

By closure of the Prototethyan oceans and matching of the Yangtze, Xianggan and Cathaysia Microcontinents, a subcontinent, sometimes named the Huanan (South China), brought into being in late Early Paleozoic. The Indosinian Microcontinent might be partly collided the subcontinent, and separated from the Yangtze affinity by the opening of the Song Ma and Mojiang oceans in Late Paleozoic. So far, the Early Paleozoic high pressure metamorphism has not been observed in the middle-south area of China, which is different from the Song Da belt where the glaucophane schist yields an age of 455 Ma[35].

PALEOTETHYAN EVOLUTION

The Paleotethyan evolution, an example from west Yunnan, southeast Xizang and southwest Sichuan, experienced a complicated Wilson cycle. Four steps can be distinguished as follows.

The Oceanic Crust Forming
According to the dating of 385 Ma from gabbro in the Tongchangjie ophiolite[11] and the REE content and Ce abundance of chert in Menglian[15], the Bitu-Changning-Menglian Oceanic Basin can be considered to be created during Mid-Late Devonian. The Carboniferous seems to be the bloom of oceanic crust occurring, during that time the ocean bacame a paleobiogeographical demarcation of the Gondwana and the Yangtze domains. Small blocks and oceanic islands made this oceanic basin to appear as an archipelagic ocean (Figs.3a, 4).

Based on the discoveries that the the Devonian is deposited unconformably on the underlying strata bordering Sichuan and Xizang[7] and the Silurian on both sides of the Jinshajiang Oceanic Basin can be well correlated[39], it is thought that the Jinshajiang Oceanic Basin was evolved from the rifting of a unified continental basement. The extention began in Early Devonian and further developed in Late Devonian-Early Carboniferous, accompanied with violent volcanism. The oceanic crust is inferred to have existed in Late Carboniferous.

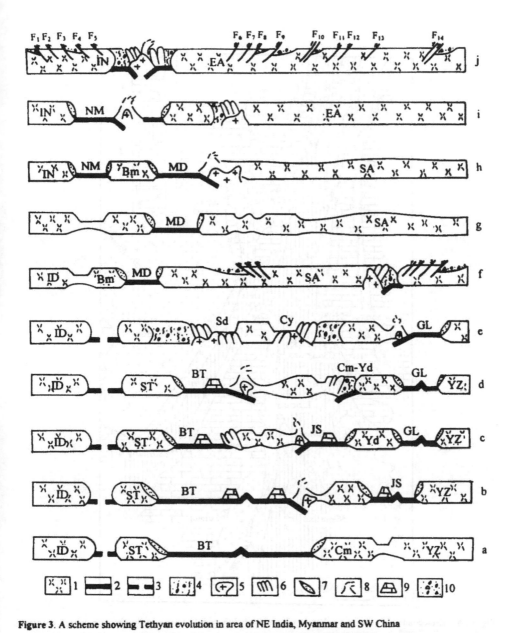

Figure 3. A scheme showing Tethyan evolution in area of NE India, Myanmar and SW China
1. Continental crust; 2. Oceanic crust; 3. Oceanic crust inferred; 4. Ophiolitic melange; 5. Granitoid; 6. Accretionary complex; 7. Passive continental margin deposits; 8. Volcanic arc; 9. Oceanic island or seamount; 10. Molasse; a. Devonian; b. Early Permian; c. Late Permian-Early Triassic; d. The latest Middle Triassic-earliest Late Triassic; e. Mid Late Triassic; f. The latest Triassic-earliest Jurassic; g. Middle Jurassic; h. Early Cretaceous; i. Late Cretaceous; j.Tertiary; Name of tectonic units: Bm. the Burma Block; EA. the Eurasia plate; ID. the Indian Gondwana; IN. the India plate; MD. the Mandalay ocean; NM. the Naga-Manipur ocean; SA. the Southeast Asian Subcontinent; for others, see Figure 2; Name of the thrusts: F_1 Naga; F_2 Disang; F_3 Tipu; F_4 Zepuhu; F_5 Nimi; F_6 Nujiang; F_7 Yezi; F_8 Yunling; F_9 Ludian; F_{10} Heqin; F_{11} Chenghai; F_{12} Yupaojiang; F_{13} Luzhijiang; F_{14} Xiaojiang

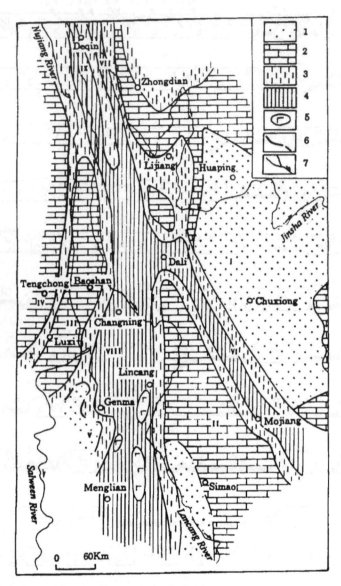

Figure 4. Paleotectonic map of western Yunnan during Carboniferous. Positions on the map are present geographic ones, the same as in Figures. 5, 6 and 7.

1. Upwarped and denued district; 2. Stable block; 3. Continental slope (passive continental margin); 4. Oceanic Basin; 5. Oceanic island; 6. Boundary of tectonic units (dashed line for inference); 7. River; I. The Yangtze Microcontinent; II. The Simao Block; III-V. The Shan-Thai Microcontinent: III. The Baoshan Block; IV. The East Kachin-Tengchong Block; V. The Cangyuan Paleo-island; VI. The Mojiang Oceanic Basin; VII. The Jinshajiang Oceanic Basin; VIII. The Changning-Menglian Oceanic Basin; IX. The Bitu Oceanic Basin; X. The Luxi Trough

The continuous Silurian-Devonian deposits near Mojiang are recently reported by Chen Tingen (see Zhong Dalai et al.[103]), which suggests the Mojiang basin may inherit partly from a Silurian trough. The oceanic crust may arise in Early Carbonifeous, which is evidenced by the dating of 339 Ma from gabbro in the Shuanggou ophiolite [11].

The contributions of paleomagnetic study are as follows. The paleolatitude of the Emeishan basalt, in an age of early Late Permian and along the recently west margin of the Yangtze Microcontinent, is 3.8 degrees south latitude[34]. The Chaya-Mangkang Block was situated near the equator during Late Permian (5 degrees north latitude for the Tuoba coal measures in Mangkang, east Xizang, see Huang Kainian et al.[33]). The paleolatitude of the Woniusi basalt from Baoshan being 42 degrees south latitude [32] suggests that the Shan-Thai Microcontinent was situated in southern hemisphere in Early Permian. In addition, the North Xizang Microcontinent was located in mid latitude position of southern hemisphere during Carbono-Permian, with Carboniferous latitude of the Qiangtang Block (sample from Yaxico) and Early Permian one of the Lhasa Block (sample from north of Lhasa) being 20 and 23.5 degrees south latitude respectively[16, 42]. The paleomagnetic results and paleobiogeographical domains indicate the Paleotethys must have been a wide ocean, with width of thousands kilometers. This opinion is also supported by sedimento-geochemical study of the Carbono-Permian silicolites[17].

The Oceanic Crust Subducting

The Bitu-Changning-Menglian Oceanic Basin partly subducted eastwards in the latest Carboniferous and Early Permian, but not closed (Figs.3b, 5). The accompanying high presure metamorphism yields an age of 279 Ma (plateau age of crossite, see Zhao Jing et al.[102]), and the granitism of 279-297 Ma (whole rock Rb-Sr isochron age, see Cong Bolin et al.[11]). The Jinshajiang-Mojiang oceanic belt subducted westwards in Late Permian, and the closure time varies in different sections. The Mojiang ocean disappeared by the end of Permian. The subduction in the southern sector of Jinshajiang Oceanic Basin might extend to Early Triassic, then a relic oceanic basin occurred till late Middle Triassic. The northern sector of the ocean kept on developing in early Late Triassic, which was evidenced by the existing active continental margin[30]. So far, no high pressure metamorphic minerals have been discovered along the southern sector of Jinshajiang geosuture zone. Owing to the westward subduction of the Jinshajiang-Mojiang oceanic belt, a large-scale rifting occurs in the west marginal area of the Yangtze Microcontinent, which is recorded by the Yanyuan-Lijiang Pericontinental Rift[75] and the Panzhihua-Xichang Rift System[10] in the region bordering Sichuan and Yunnan; also, it leads to the opening of the Ganzi-Litang Oceanic Basin in western Sichuan (Figs.3c, 6).

The Relic Oceans

Although extention and continental rifting occurred in Early-Middle Triassic, the oceanic basins inherited the Permian ones; Wu Genyao and Yano[83] put forward that the Triassic Tethys should be the last period of Paleotethyan evolution, analogous to the Mediterranean stage of Wilson cycle, instead of the opening of Neotethys. Unlike the wide archipelago pattern of Late Paleozoic, the Triassic Paleotethys of this stage was made up of several relatively small isolated oceanic basins (Fig.7), where there might occur the oceanic islands, for example, the Laochang Volcanics in the Lancang Oceanic Basin, a relic one after partly subduction of the Changning-Menglian Oceanic Basin[77].

The Oceanic Closure and Continent (arc)-Continent (arc) Collision

The southern sector of the Jinshajiang ocean was closed in late Middle Triassic, with the molasse in an age of Carnian. Following the closure the Yidun Block collided with the Chaya-Mangkang Block, giving rise to the development of collision-type granitoids and syncollisional volcanics in the latter (Figs. 2, 3d). The granitoids are represented by the Ludian and Jiaren rock masses of 234.3 Ma (whole rock Rb-Sr isochron age) and 228.7 Ma (biotite K-Ar age) respectively[7]. To the west of the

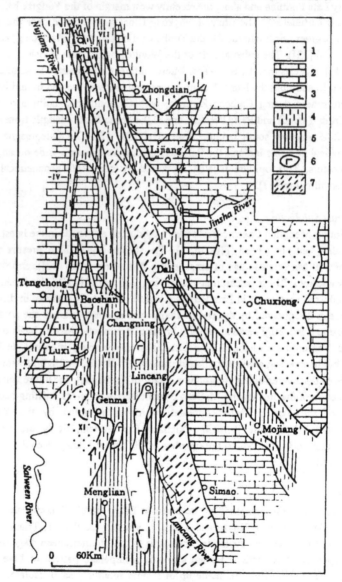

Figure 5. Paleotectonic Map of western Yunnan during Early Permian

1-2. The Same as 1-2 in Figure 4; 3. Intracontinental rift; 4-6. The same as 3-5 in Figure 4; 7. Active continental margin; I-IV. The same as I-IV in Figure 4; V. The Chaya-Mangkang Block; VI-X. The same as VI-X in Figure 4

granitoids there develops the east branch of the Northwestern Yunnan Rift System[83], filled with the bimodal volcanic rocks. The lower part of the volcanics is acidic, formed during collision by anatexis of violent crustal compression, with materials sourced from continental crust. The upper part is basalt sandwiched with rhyolite, the volcanic materials coming from a depleted mantle[101]. The rift development is related to the de-rooting and lithospheric delamination in the collisional orogenic area[84]. In the Yidun Block, the underthrusting plate, occurs the foreland fold-and-thrust belt, consisting of a group of reversed faults which are thrusted successively eastwards and control a molasse basin developing.

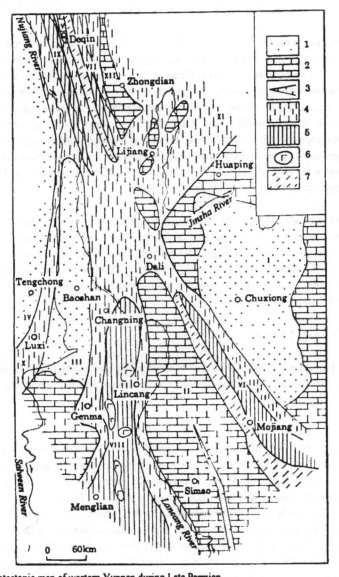

Figure 6. Paleotectonic map of western Yunnan during Late Permian

1-7. The same as 1-7 in Figure 5; I-X. The same as I-X in Figure 5; XI. The Yanyuan-Lijiang Pericontinental Rift; XII. The Yidun Block

The closure time of the southern sector of the Bitu Oceanic basin seems relatively earlier. It was subducted eastwards since Middle Triassic, and the Shan-Thai Microcontinent came into collision with the Chaya-Mangkang Block in the latest Middle Triassic and early Late Triassic. Three belts, namely, the belts of ophiolitic melange, collision-type granitoids and syncollisional volcanics are closely related spatially and distributed parallely in the block (Fig. 2, 3e). The rock mass at Baimang Snow Mt. of 220 Ma and 223.1 Ma (biotite K-Ar age[7]) represents the collision-type granitoids. The mantle-sourced basalt records the west branch of the Northwestern Yunnan Rift System, whose extension seems more limited than the east branch, since the Lancang ocean and the northern sector of the Bitu Oceanic Basin remain on occurring.

All of the Paleotethyan oceanic basins were finally closed in Mid Norian, followed with the Jurassic Tethyan oceans opened, which anounced the beginning of another tectonic cycle: the Neotethyan evolution (Fig. 3f). The Bitu and Lancang Oceanic Basins are subducted towards east and the Ganzi-Litang one towards west. During the period of violent continent-continent collision, the obducting-nappe structuring is well developed in the amalgamated Chaya-Mangkang Yidun Blocks which is an overthrusting plate[82]; and the foreland thrusting occurs in previously passive continent marginal regions of the Shan-Thai and Yangtze Microcontinents respectively. Owing to the crustal wedges stacking up and the upper crust thickening, the orogenic area are uplifted and eroded for a long time, till early Middle Jurassic. Since Middle Jurassic a super-large red basin brought into being in the amalgamated subcontinent (Fig.3g).

The collision and amalgamation of the Huanan Subcontinent with the Indosinian, Zheming and Shan-Thai Microcontinents, as well as the Qiangtang Block and the median masses in the Paleotethyan oceans, created the Southeast Asia Subcontinent. The Nanhai Microcontinent might become a part of the subcontinent by the end of Triassic. In addition, the Yangtze collided and welded with the North China Microcontinents in the Indosinian orogenesis[76], and the China-SE Asia Subcontinent brought into being.

NEOTETHYAN EVOLUTION

The Neotethyan evoluiton in South China and its environs can be divided into two sub-cycles, and correspondingly, the Yanshanian orogenesis in Juro-Cretaceous and the Himalayan orogenesis in Tertiary are called respectively.

The Yanshanian Orogenesis

In Southwest China and Myanmar, following closure of Paleotethys, the northern margin of Indian Gondwana may be rifted, which resulted in the Bangonghu-Dengqen Oceanic Basin between the Qiangtang and Lhasa Blocks and the Mandalay Oceanic Basin between the Shan-Thai Microcontinent and the Burma Block to take shapes in the age of the latest Triassic and Early Jurassic (Fig.3f). The idea is further confirmed by the paleomagnetic measurements which suggest that the Shai-Thai Microcontinent was moved to a low latitude position of the northern hemisphere in Early Triassic[4], and the Qiangtang Block was drifted northwards from 20 degrees south latitude in Carboniferous to 25.5 degrees north latitude in Triassic[42]. Yu Guangming et al.[92] thought the Bangonghu-Dengqen ocean, occurring in Jurassic, consisting of some pull-apart basins controlled by strike-slipping of the Bangonghu-Dengqen fracture. So far the oceanic basins' subduction vergence is unknown[95]. The Mandalay Oceanic Basin might subduct eastwards in Early Cretaceous. The related granitoids in the Tenasserim belt, east Myanmar, yield Rb-Sr ages of 143-130 Ma and 120-110 Ma[2]. The Cretaceous granitoids in Tengchong, the westmost Yunnan, can be divided into two generations. The earlier one, in an age of 143-100 Ma, is related to arc-magmatism; and the later one, in an age of 96-78 Ma, is characterized with tinny adamellite[7, 48, 99]. To the east of the magmatic arc occurred a backarc continental basin in west and central part of Yunnan Province during Early Cretaceous[90].

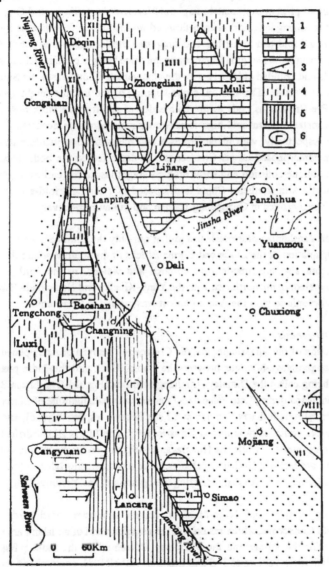

Figure 7. Paleotectonic map of western Yunnan during Anisian, Middle Triassic
1-6. The same as 1-6 in Figure 6; I. The East Kachin-Tengchong Block; II. The Luxi Trough; III. The Baoshan Platform; IV. The Cangyuan Platform; V. The Shanglan Aulacogen; VI. The Simao Platform; VII. The Niushang Aulacogen; VIII. The Gejiu Platform; IX. The Yanyuan-Lijiang Platform; X. The Lancang (relic Changning-Menglian) Oceanic Basin; XI. The Bitu Oceanic Basin; XII. The Jinshajiang Oceanic Basin; XIII. The Songpan-Ganzi Trough

The area of southeast China is involved in the Yanshanides. The Early Jurassic bimodal volcanic rocks and deep marine sediments with ammonite fauna of Tethyan affinity in region bordering Guangdong and Fujian might represent an aulacogen, which linked up with the Neotethyan ocean in the south, whose ophiolites were preserved in the Meratus Mountains, southeastern Borneo, as well as the Darvel Bay area (Sabah, Palawan and Tawi-Tawi, see Hutchison[35]). The oceanic crust might subduct since the latest Early Jurassic, which was responsible for the Jurassic episode of rifting and forming of present continental margin of Northern Australia-central Guinea[1]. The W-E striking fractures in Guangdong and Hainan thrusted northwards in Yanshanian movement, which represented a foreland fold-and-thrust belt and suggested the Jurassic ocean subducted southwards. With the oceanic closure, the Donghai Microcontinent, once being a continental massif in the Tethys, moved northwards and collided obliquely with the amalgamated Southeast Asia Subcontinent along the Changle-Nan'ao fracture. The sinistral strike-slipping of the fracture might go on in Mid-Late Jurassic, so the Pre-Cretaceous beds in the both sides of the fracture were metamorphosed. The medium pressure metamorphism and high temperature metamorphism occurred in the lower wall and hanging one respectively, with metamorphism and migmatization yielding an ages of 127-90 Ma[78]. The unmetamorphosed Lower Cretaceous was deposited on the both sides of the fracture, implying that the oblique collision and shear orogeny might last to the earliest Cretaceous. In short words, the Yanshanides in southeast China is composed of a foreland fold-and-thrust belt and an oblique collision-shear orogen in Guangdong-Hainan and Fujian respectively.

The importance of Yanshanian orogenesis lies in that not only the China-SE Asia Subcontinent was enlarged by amalgamation of the Lhasa and Burma Blocks in the southwest and the Donghai Block in the southeast, but also the subcontinent collided and welded, in the north, with the Siberia Craton[60] and accreted, in the northeast, with the Yanji-Khanka and Nadanhada-Bikin terrenes[20]. As a result, an embryonic form of the Asia continent was created in Cretaceous.

The Himalayan Orogenesis
Following subduction of the Jurassic oceans, the Cretaceous Tethyan oceans opened along a line of Indus-Yarlung Zangbo-Naga-Manipur between the Lhasa Block and the India plate, the northern part of the latter being called the Himalaya (or Jiangzi) Block (Fig.3h). The Lhasa Block was still a part of Indian Gondwana during Late Paleozoic-Triassic, with paleolatitude of Lhasa being 23.5 and 21.4 degrees south latitude in Early Permian and Late Triassic respectively[16], and the Himalayan district in 24 degrees south latitude during Triassic[109]. The Lhasa Block drifted northwards to equator ($1°N$) in Late Jurassic, while the Himalayan district seemed to be in situ, with paleolatitude of Jiangzi $23°S$ in Late Jurassic[109]. The paleomagnetic measurements later on (cf. Wu Gongjian et al.[85]) supported further these results, indicating the opening of the Yarlung-Zangbo ocean.

The Indus-Yarlung Zangbo and Naga-Manipur oceans subducted northwards and eastwards respectively in Late Cretaceous, which resulted in the establishment of an island arc system, namely, the Gangdise Arc for the former and the Mandelay Volcanics for the latter (Fig.3i). The Late Crateceous paleolatitudes, Yadong at $3°S$ and Lhasa at $7.3°N$[108], comprehend the ocean did not disappear. The Naga-Manipur ocean closed in Eocene, followed collision of the India plate with the island arc mentioned above, then the India plate welded to the island arc colliding the Burma Block, which caused the thrusts in northeast India to migrate in a northwesrterly direction (Fig. 3j, see Bhattacharjee[3]). The thrusts in Yunnan were migrated eastwards in Eocene and Oligocene[79], which might suggest the back arc basin, to the east of the Mandalay magmatic arc, to be subducted towards west (Fig. 3j).

Late Tertiary is an important period for intracontinental deformation. The large-scale strike-slipping occurred in Miocene along the main fractures, such as the Sagaing[23], Nujiang[14, 72], Wuliangshan-Yingpanshan[25] and Ailaoshan fracture[62, 106]. As a result of dextral strike-slipping of the former two and sinistral one of the latter two, the area sandwiched in between the Sagaing and Ailaoshan fractures tectonically escaped towards southeast by south[46, 82]. In addition, owing to arising of the Qinghai-Xizang plateau in Pliocene and Quaternary, the bilateral thrusting occurred in the marginal areas of the plateau[79]. The collision between the India and Eurasia plates and intracontinental deformation may have continued to present time, and the India plate is still, tectonically, an independent unit with respect to the Eurasia plate, although geographically it has become one part of the Asia continent.

CONCLUDING REMARKS

The Indosinian orogenesis, related to the Paleotethyan evolution, laid a foundation of tectonic features in the South China area. The Yanshanian and Himalayan orogenies, related to the Neotethyan evolution, created the Qinghai-Xizang Plateau and the continental margin of the southeast coast area of China, and reworked the Indosinian features. In other words, Asia continent is a composite one, resulted from amalgamation of a number of microcontinents and blocks which were originated from Gondwana and were once separated by Tethyan oceans or troughs. Tethyan evolution, which plays a key role in arising of today's tectonic framework of South China and its environs, is the crux for understanding the regional geohistory in the light of mobilism and global tectonics. The Tethysides is a very complicated orogenic zone, including subduction, collision and shearing orogenic movements. Owing to multigenerations of reformation, South China and its environs is, in reality, a superlarge Tethyside orogenic collage.

Tethyan evolution is in itself a process of Gondwana dispersion and Asian accretion. The closure of Prototethys caused the Yangtze, Xianggan and Cathaysia Microcontinents, which might be originated from different parts of Gondwanaland, to amalgamate a subcontinent, the Yangtze affinity or the Huanan Subcontinent. The Indosinian, Shan-Thai, Zheming, North China and Donghai Microcontinents might be dispersed from the Australian Gondwana in Sinian and early period of Paleozoic[8], and grouped to be Australian Gondwana affinity. The former four matching the Huanan Subcontinent by closure of the Paleotethys created the China-SE Asia Subcontinent in Triassic. The Nanhai Microcontinent remained to be a part of Australian continent during Cambrian[76], its faunas in Early Carboniferous was of Gondwana affinity[68]. It is named the Australian Gondwana in the paper. The Qiangtang and Lhasa Blocks belonged to the Indian Gondwana since the both were a part of India continent during Paleozoic. The Indosinian and Yanshanian orogenies, such as mentioned above, played a key role in arising of the Asian continent. The subduction of Cretaceous Tethyan oceans and collision of the India and Eurasia plates lead to the present Asia continent to take shape.

The geohistory of South China and its environs during Sinian and Phanerozoic time can be summarized as three major tectonic cysles. Every cycles is initiated with continental rifting and Gondwana disintegration, and terminated with continental welding and accretion which resulted in Asian enlargement. The three cycles are well corresponding to geotectonic evolutions of Prototethys,

Paleotethys and Neotethys respectively, and the latter can be divided into two subcycles. A tectonic differentiation between the west and east parts of the area since Jurassic should be emphasized. The kinematics of subduction and collision in the west part, namely, the Qinghai-Xizang Plateau area, seems similar to what occurred in Triassic; while in the southeast coast region of China and its neibouring areas, sinistral strike-slipping of a group of NE-SW striking fracture, for example, the Changle-Nan'ao fracture, dominate the regional geological development. On the one hand, the Qinghai-Xizang Plateau area has compressed and uplifted violently, with the main tectonic line of NWW-SEE direction; on the other hand, the east China area, with the main tectonic line of NNE-SSW striking, has transtensioned; and the region sandwiched in between the both, namely, between the Sagaing and Ailaoshan fractures, has tectonically escaped. The South China and its environs provides a best example for such an idea: Tethys must have been a global-scale realm of transform tectonics.

Acknowledgement

The project is financially supported by the National Natural Science Foundation of China, as well as the State Scientific and Technological Commission. The author is sincerely grateful to Profs. Liu Hefu, Zhong Dalai and Zhang Shaozhong for their invaluable suggestions to the manuscript, and to Profs. Liu Benpei, Cong Bolin, Zhang Qi and Pan Yusheng for their helpful discussions.

REFERENCES

1. M.G. Audley-Charles. Evolution of the southern margin of Tethys (North Australian region) from early Permian to late Cretaceous. In: *Gondwana and Tethys*. M.G. Audley-Charles and A. Hallam (Eds.). pp.79-100. Oxford University Press, Oxford (1988).

2. F. Bender. *Geology of Burma*. Gebruder Borntraeger, Berlin (1983).

3. C.C. Bhattacharjee. The ophiolites of northeast India - a dubduction zone ophiolite complex of   the Indo-Burman orogenic belt. *Tectonophysics* 191, 213-222 (1991).

4. S. Bunopas and P. Vella. Tectonic and geologic evolution of Thailand. In: *Proceedings of the workshop on stratigraphic correlation of Thailand and Malaysia*. Geological Society of Thailand and Geological Society of Malaysia (Eds.). pp.307-322. Haad Yai, Thailand (1983).

5. Bureau of Geology and Mineral Resources of Fujian Province. *Regional Geology of Fujian Province*. Geological Publishing House, Beijing (1985) (in Chinese).

6. Bureau of Geology and Mineral Resources of Xizang Autonomous Region. *Regional geology of Xizang Autonomous region*. Geological Publishing House, Beijing (1993) (in Chinese).

7. Bureau of Geology and Mineral Resources of Yunnan Province. *Regional geology of Yunnan Province*. Geological Publishing House, Beijing (1990) (in Chinese).

8. C.F. Burrett and B. Stait. China and Southeast Asia as a part of the Tethyan margin of Cambro-Ordovician Gondwanaland. In: *Shallow Tethys 2*. K. Mckenzie (Ed.). pp.65-77. Balkema, Rotterdam (1987).

9. Chang Chengfa and Bian Qiantao. The multi-branched orogenic complex of the Indosinides: nature and evolution of Paleo-Tethys. In: *Advances in Geoscience (2)*. Institute of Geology, Academia Sinica (Ed.). pp.1-8. China Ocean Press, Beijing (1992).

10. Chinese Members of Co-operative Geology Group of China and Japan in the Pan-Xi Region. *Petrotectonics of Panzhihua-Xichang region, Sichuan Province, China*. China Ocean Press, Beijing (1988).

11. Cong Bolin, Wu Genyao, Zhang Qi et al. Petrotectonic evolution of Paleotethys in western Yunnan, China, *Science*

in China (Series B) **37**, 1016-1024 (1994).

12. Deng Wanming. Preliminary investigation into the mafic and ultramafic rocks in the western Kunlun -Karakorum, *Journal of Natural Resources* **4**, 204-211 (1989) (in Chinese).

13. Deng Wanming. Genetic environment of the Kudi ophiolite suite and tectonic significance, Xinjiang, China, *Mem. Lithosph. Tect. Evol. Res.* (1), 82-88 (1993).

14. Ding Lin and Zhong Dalai. Characteristics of deformation in the Gaoligong strike-slip fault, western Yunnan, China. In: *Advances in geoscience (2)*. Institute of Geology, Acdamia Sinica (Ed.). pp.82-92. China Ocean Press, Beijing (1992).

15. Ding Lin, Zhong Dalai and Li Hongsheng. Rare earth elements and stable isotopes study in chert from the Eastern Paleotethys, western Yunnan, China In: *Ann. Rep. Lab. Lithosphere Tectonic Evolution, Inst. Geol., Chinese Academy of Sciences (1993-1994)*. Lab. LTE (ed.). pp.156-160. Seismological Press, Beijing (1994).

16. Dong Xuebin, Wang Zhongmin, Tan Chengze et al. New paleomagnetic results from Yadong-Golmud geoscience transect and a preliminary study on the model of terrenes evolution in Qinghai-Xizang Plateau, *Bull. Chinese Academy Geological Sciences* **21**, 139-148 (1990) (in Chinese).

17. Du Yue and Wu Haoruo. The geochemical study on the Carboniferous-Permian siliceous rocks of Menglian, Yunnan Province, *Acta Petrologica Sinica* **9** (suppl.), 195-199 (1993) (in Chinese).

18. Duan Xinhua and Zhao Hong. The Ailaoshan-Tengtiaohe Fracture - the subduction zone of an ancient plate, *Acta Geologica Sinica* **55**, 258-166 (1981) (in Chinese).

19. Fang Runsen. The early Permian brachiopoda from Xiao Xinzhai of Gengma, Yunnan and its geological significance. In: *Contribution to the geology of the Qinghai-Xizang (Tibet) Plateau.* CGXQP Editorial Committee (ed.). pp.93-119. Geological Publishing House, 11, Beijing (1983) (in Chinese).

20. Ge Xiaohong, Liu Xianwen and Liu Ping. Mesozoic collision tectonics in eastern Jilin and Heilongjiang Province, Northeast China: A coexistence of two tectonic regimes. In: *Abstract volume, IGCP 321 fourth international symposium and field excursion.* M. Chao and J.H. Kim (Eds.). pp.33-35. Harn Lim Printing Co., Seoul (1994).

21. Hao Taiping. Sm-Nd isotopic ages of Proterozoic metamorphic rocks from the middle sector of the Jinsha River, *Geological Review* **39**, 52-56 (1993) (in Chinese).

22. He Haiqing, Li Jiliang and Xiao Wenjiao. Carbonate turbidite of the Lower Triassic Zhengtang Formation in Zhejiang Province, China. In: *Ann. Rep. Lab. Lithosphere Tectonic Evolution, Inst. Geol., Chinese Academy of Sciences (1993-1994)*. Lab. LTE (ed.). pp.152-155. Seismological Press, Beijing (1994).

23. Hla Maung. Transcurrent movement in the Burma-Andaman Sea region, *Geology* **15**, 911-912 (1987).

24. Hou Liwei, Fu Deming, Dai Bingchun and Xiao Yi. The principal characteristics of magmatic activity in East Xizang and West Sichuan and their relationship with plate tectonics and endogenic ore deposits. In: *Contribution to the geology of the Qinghai-Xizang (Tibet) Plateau.* CGXQP Editorial Committee (Ed.). pp.69-105. Geological Publishing House, 13, Beijing (1983) (in Chinese).

25. Hou Quanlin and Zhong Dalai. The deformation and metamorphism in the Wuliangshan ductile shear zone in western Yunnan, China, *Mem. Lithosph. Tect. Evol. Res.* (1), 24-29 (1993).

26. Hou Quanlin, Li Peijun and Li Jiliang. Turbidite and contourite of the Lower Triassic Xikou Formation in southwestern Fujian. In: *Ann. Rep. Lab. Lithosphere Tectonic Evolution, Inst. Geol., Chinese Academy of Sciences (1993-1994)*. Lab. LTE (Ed.). pp. 149-151. Seismological Press, Beijing (1994).

27. Hou Quanlin, Li Peijun and Li Jiliang. *Foreland fold-thrust belt in southwestern Fujian, China.* Geological Publishing House, Beijing (1995) (in Chinese).

28. Hu Chengzu. Evolution of Xizang (Tibet) Plate in view of historical development of Yarlung Zangbo River suture zone and Chabu-Chasang Rift. In: *Contribution to the geology of the Qinghai-Xizang (Tibet) Plateau.* CGXQP Editorial Committee (Ed.). pp.111-121. Geological Publishing House, 9, Beijing (1986) (in Chinese).

29. Hu Shihua, Luo Daixi and Li Kaiyuan. Triassic sedimentary facies in East Xizang - West Sichuan and their tectonic significance. In: *Contribution to the geology of the Qinghai-Xizang (Tibet) Plateau.* CGXQP Editorial

Committee (Ed.). pp.107-128. Geological Publishing House, 13, Beijing (1983) (in Chinese)

30. Hu Xiangsheng, Mo Xuanxue and Fan Li. Jiangda ancient trench-arc-basin system, Xizang (Tibet): evidences from volcanic rocks. In: *Contribution to the geology of the Qinghai-Xizang (Tibet) Plateau.* CGXQP Editorial Committee (Ed.). pp.1-15. Geological Publishing House, 20, Beijing (1990) (in Chinese).

31. Huang Jiqing and Chen Bingwei. *The evolution of the Tethys in China and adjacent regions.* Geological Publishing House, Beijing (1987).

32. K. Huang and N.D. Opdyke. Paleomagnetic results from the upper Carboniferous of the Shan-Thai-Malay Block of western Yunnan, China, *Tectonophysics* 192, 333-344 (1991).

33. K. Huang, N.D. Opdyke, X. Peng and J. Li. Paleomagnetic results from the upper Permian of the eastern Qiangtang terrene of Tibet and their tectonic implications, *Earth Planet. Sci. Lett.* 111, 1-10 (1992).

34. Huang Kainian, N.D. Opdyke, D.V. Kent et al. Some new paleomagnetic results from the Permian Emeishan basalt, *Chinese Science Bulletin* 31, 133-137 (1986) (in Chinese).

35. C.S. Hutchison. *Geological evolution of South-East Asia.* Clarendon Press, Oxford (1989).

36. Jin Xiaochi. *Sedimentary and paleogeographic significance of Permo-Carboniferous sequences in Western Yunnan, China.* Geol. Inst. Univ. Koln Sonderveroeffentlichungen, Num.99 (1994).

37. Lan Chaohua, Sun Cheng, Fan Jiancai and Fang Runsen. Carboniferous and Permian stratigraphy of the Zhenkang and Luxi region in Western Yunnan. In: *Contribution to the geology of the Qinghai-Xizang (Tibet) Plateau.* CGXQP Editorial Committee (Ed.). pp.79-91. Geological Publishing House, 11, Beijing (1983) (in Chinese).

38. Li Cai. The Longmucuo-Shuanghu-Lancangjiang plate suture and northern boundary of distribution of Gondwana facies of Permo-Carboniferous system in northern Xizang, China, *Journal of Changchun College of Geology* 17, 155-166 (1987) (in Chinese).

39. Li Xinzhen, Liu Zengqian, Pan Guitang, Luo Jianning, Wang Zheng and Zheng Lailing. The Dividing and evolution of tectonic unit of Sanjiang, southwest China, *Bull. Chengdu Inst. Geo. M. R., Chinese Acad. Geol. Sci.* (13), 1-20 (1991) (in Chinese).

40. Li Yuejun, Hao Jie, Li Hongsheng and Yang Meifang. The Chabu-Chasang zone in northern Xizang, *Scientia Geologica Sinica* 3, 151-160 (1994).

41. Liang Dingyi, Nie Zetong, Guo Tieying et al. Permo-Carboniferous Gondwana-Tethys facies in Southern Karakorum Ali, Xizang (Tibet), *Earth Science* 8, 9-27 (1983) (in Chinese).

42. Lin Jinglu. Advances in the paleomagnetic study of China. In: *Advances in Geoscience (1).* Wang Sijing (Ed.). pp.20-33. China Ocean Press, Beijing (1989).

43. Liu Benpei. On the tectono-palaeogeographical development of South China in the Hercynian-Indosinian stage. In: *Tectonic history of the ancient continental margins of South China.* Wang Hongzhen, Yang Weiran and Liu Benpei (Eds) pp.65-77. China University of Geosciences Press, Wuhan (1986) (in Chinese).

44. Liu Benpei and Cui Xinsheng. Discovery of *Eurydesma*-fauna from Rutog, Northwest Xizang (Tibet), and its biogeographic signification, *Earth Science* 8, 79-92 (1983) (in Chinese).

45. Liu Hefu, Liang Huishe, Cai Liguo and Shen Fei. The evolution of the Palaeo-Tethys and the formation and deformation of the superimposed basins in western Sichuan and western Yunnan provinces. In: *Tectonopalaeogeography and palaeobiogeography of China and adjacent regions.* Wang Hongzhen et al. (Eds.).pp.89-108. China University of Geosciences Press, Wuhan (1990) (in Chinese).

46. Liu Sanyin, Zhong Dalai and Wu Genyao. Formation and evolution of Late Tertiary coal-bearing basin in southwest Yunnan, *Journal of China Coal Society* 20, 351-355 (1995) (in Chinese).

47. Liu Zengqian et al. A preliminary study on the north boundary and the evolution of Gondwana and Tethys in light of the new data on Qinghai-Xizang (Tibet) Plateau. In: *Contribution to the geology of the Qinghai-Xizang (Tibet) Plateau.* CGQXP Editorial Committee (Ed.). pp.11-24. Geological Publishing House, 12, Beijing (1983) (in Chinese).

48. Liu Zengqian, Li Xingzhen, Ye Qingtong, Luo Jianning , Shen Ganfu et al. *Division of tectono-magmatic zones*

and the distribution of deposits in the Sanjiang Area. Geological Publishing House, Beijing (1993) (in Chinese).

49. Luo Junlie. Evolution and basic characteristics of Tethys orogenic zone, *Yunnan Geology* **9**, 247-290 (1990) (in Chinese).

50. Luo Yaonan. Panzhihua-Xichang paleo-rift zone, China. In: *Contribution to Panzhihua-Xichang Rift, China.* Zhang Yunxiang (Ed.). pp.1-25. Geological Publishing House, 1, Beijing (1985) (in Chinese).

51. A.S. Macdonald and S.M. Barr. Nan River mafic-ultramafic belts, northern Thailand: Geochemistry and tectonic significance, *Geol. Soc. Malaysia Bull.* **17**, 209-224 (1984).

52. Mo Xuanxue, Lu Fengxiang, Shen Shangyue, Zhu Qinwen, Hou Zengqian et al. *Sanjiang Tethyan volcanism and related mineralization.* Geological Publishing House, Beijing (1993) (in Chinese).

53. Pan Guitang, Zheng Haixiang, Xu Yuerong, Wang Peisheng and Jiao Shupei. A preliminary study on Bangong Co-Nujiang suture. In: *Contribution to the geology of the Qinghai-Xizang (Tibet) Platear.* CGXQP Editorial Committee (Ed.). pp.229-242. Geological Publishing House, 12, Beijing (1983) (in Chinese).

54. Pan Yusheng. Ophiolite suite was discovered in Tongtian River, Qinghai province, *Seismology and Geology* **6**, 44 (1984) (in Chinese).

55. Pan Yusheng. Proto-Tethys track in western Kunlun, China, *Mem. Lithosph. Tect. Evol. Res.* (1), 30-35 (1993).

56. Peng Xinjie and Hu Changshou. Tectonic evolution of the Sanjiang belt in eastern Tibet, *Regional Geology of China* (2), 140-148 (1993) (in Chinese).

57. Peng Xinjie and Luo Wanlin. The discovery of the glaucophane schist zone in the southern section of the Lancang Jiang, western Yunnan and its tectonic significance, *Regional Geology of China* (2), 69-75 (1982) (in Chinese).

58. Qiu Yuanxi. Link between China continent and Southeast Asia in geological evolution. In: *Abstract volume, IGCP 321 fourth international symposium and field excursion.* M. Chao and J.H. Kim (Eds.). pp.101-102. Ham Lim Printing Co., Seoul (1994).

59. Ren Jishun. On the geotectonics of southern China, *Acta Geologica Sinica* **64**, 275-288 (1990) (in Chinese).

60. Ren Jishun, Chen Tingyu, Niu Baogui, Liu Zhigang and Liu Fengren. *Evolution of the continental lithosphere and metallogeny in eastern China and adjacent area.* Science Press, Beijing (1990) (in Chinese).

61. Ren Shengli, Li Jiliang, Hao Ji, Zhang Jianbo and Hu Xiongjian. Xiongshang diabase swarm, North Fujian: Petrochemistry and tectonic implications. In: *Ann. Rep. Lab. Lithosphere Tectonic Evolution, Inst. Geol., Chinese Academy of Sciences (1993-1994).* Lab. LTE (Ed.). pp.88-90. Seismological Press, Beijing (1994).

62. U. Schaarer, P. Tapponnier, R. Lacassin, P.H. Leloup, Z. Dalai and J. Shaocheng. Intraplate tectonics in Asia: a precise age for large-scale Miocene movement along the Ailao Shan-Red River shear zone, China, *Earth Planet. Sci. Lett.* **97**, 65-77 (1990).

63. Sha Shaoli. On the glaucophane schist of the join between the Zhongdian, Yunnan and Muli, Sichuan, *Yunnan Geology* **7**, 82-85 (1988) (in Chinese).

64. Sun Dongli. On the Permian biogeographic boundary between Gondwana and Eurasia in Tibet, China as the eastern section of Tethys, *Palaeogeography, Palaeoclimatology, Palaeoecology* **100**, 59-77 (1993).

65 Sun Mingzhi and Xu Keqin. On the Caledonian granitoids and their geotectonic environments of South China, *Journal of Nanjing University (Earth Science Edition)* (4), 10-22 (1990) (in Chinese).

66. Sun Xiaomeng, Nie Zetong and Liang Dingyi. Determination of sedimentary environments and tectonic significance of silicolites in Jinsha River belt, NW Yunnan, *Geological Review* **41**, 174-178 (1995) (in Chinese).

67. Wang Liancheng, Li Dazhou, Zhang Qi and Zhang Kuiwu. Ophiolitic melange in Litang, Sichuan Province, *Acta Petrologica Sinica* **1**,17-27 (1985) (in Chinese).

68. Wang Hongzhen. A synopsis of the tectonic development of South China. In: *Tectonic history of the ancient continental margins in South China.* Wang Hongzhen, Yang Weiran and Liu Benpei (Eds.). pp.1-15. China University of Geosciences Press, Wuhan (1986) (in Chinese).

69. Wang Peisheng. Petrochemistry of the ophiolite-associated lavas in Deqin, Yunnan. In: *Contribution to the geology*

of the Qinghai-Xizang (Tibet) Plateau. CGQXP Editorial Committee (Ed.). pp.207-219. Geological Publishing House, **9**, Beijing (1986) (in Chinese).

70. Wang Yizhao. The characteristics and significance of Carboniferous gravel bed in Tengchong and Baoshan area, Western Yunnan. In: *Contribution to the geology of the Qinghai-Xizang (Tibet) Plateau.* CGQXP Editorial Committee (Ed.). pp.71-77. Geological Publishing House, **11**, Beijing (1983) (in Chinese).

71. Wen Shixuan. New paleontological evidence for continental drift in the Qinghai-Xizang Plateau. In: *Proceedings of the first symposium on the Qinghai-Xizang Plateau.* China Society of the Qinghai-Xizang Plateau Research (Ed.). pp.308-313. Science Press, Beijing (1992) (in Chinese).

72. Wu Genyao. Fractures and their neotectonic active modes in border area of China, Burma and Thailand, *Quaternary Sciences* (1), 28-37 (1991) (in Chinese).

73. Wu Genyao. A preliminary discussion on the stratigraphy of the Yiliu and Pingzhang Formations in Changning-Menglian area, western Yunnan, *Journal of Stratigraphy* **17**, 302-309 (1993) (in Chinese).

74. Wu Genyao. Late Paleozoic tectonic framework and Paleotethyan evolution in western Yunnan, China, *Scientia Geologica Sinica* **2**, 129-140 (1993).

75. Wu Genyao. Permian basalts in Lijiang and Jinping, west Yunnan: a comparative study and its geological significance, *Acta Petrologica Sinica* **9**(suppl.), 63-69 (1993) (in Chinese).

76. Wu Genyao. Tectonic outline of South China and its environs: with reference to Gondwana dispersion and Asian Accretion. In: *Asian accretion.* Chinese Working Group of IGCP Project 321 (Ed.). pp.17-20. Seismological Press, Beijing (1993) (in Chinese).

77. Wu Genyao. Laochang Volcanics: a record of Azores-type oceanic islands in Paleotethys, southwestern Yunnan, *Scientia Geologica Sinica* **3**, 13-23 (1994).

78. Wu Genyao. Yanshanides in southeast coast area, China. In: *Abstract volume,IGCP 321 fourth international symposium and field excursion.* M. Chao and J.H. Kim (Eds.). pp.132-134. Ham Lim Printing Co., Seoul (1994).

79. Wu Genyao. Tertiary thrusting-nappe structures in northwest Yunnan, China, *Geotectonica et Metallogenia* **18**, 331-338 (1994) (in Chinese).

80. G.Y.Wu and B.L.Cong. Tethyan evolution and SE Asian continental accretion, *Journal of Geology (Series B)* (5-6), 293-301 (1995).

81. Wu Genyao, He Fuxiang, Nie Zetong and Liang Dingyi. Paleotethyan oceanic records in Bitu, East Xizang (Tibet), China, *Abstracts, 30th international geological congress* **1**, 301 (1996).

82. Wu Genyao, Sun Xiaomeng and Zhong Dalai. *Structural geology of the central sector of the Hengduan Mountains.* Geological Publishing House, Beijing (1996).

83. Wu Genyao and T. Yano. Triassic tectonic framework and geological evolution in Yunnan and Southern Sichuan, China. In: *Advances in geoscience (2).* Institute of Geology, Academia Sinica (Ed.). pp.40-55. China Ocean Press, Beijing (1992).

84. Wu Genyao, T. Yano, Sha Shaoli and Tan Mingqing. Triassic volcanism in northwest Yunnan: delamination-related rifting in collisional orogenic belt, *Scientia Geologica Sinica* **6**(1) (1997) (in printing).

85. Wu Gongjian, Xiao Xuchang and Li Tingdong. Expose the uplift of Qinghai-Tibet Plateau: Yadong-Golmud geoscience transect in Qinghai-Tibet Plateau, China, *Earth Science* **21**, 34-40, (1996) (in Chinese).

86. Wu Haoruo, Kuang Guodun and Wang Zhongcheng. Reinterpretation of basic igneous rocks in western Guangxi and its tectonic implications, *Scientia Geologica Sinica* (3), 288-289 (1993) (in Chinese).

87. Xu Ronghua, Zhang Yuquan, Xie Yingwen, Chen Fukun, Ph. Vidal, N. Arnaud, Zhang Qiaoda and Zhao Dunming. A discovery of an Early Paleozoic tectono-magmatic belt in the northern part of west Kunlun Mountains, *Scientia Geologica Sinica* (4), 346-354 (1994) (in Chinese).

88. Yang Jiawen. Discussion on the ophiolite at Tongchangjie, Yunxian County, *Yunnan Geology* **1**, 59-71 (1982) (in Chinese).

89. Yang Zunyi, Nie Zetong, Guo Tieyin, Liang Dingyi and Zhang Yizhi. Mesozoic biotic characteristics and their

relationship to crustal evolution in the Ngari region, Tibet (Xizang), China, *Earth Science* 10, 1-8 (1985) (in Chinese).

90. T. Yano, Wu Genyao, Tan Mingqing and Sha Shaoli. Tectono-sedimentary development of backarc continental basin in Yunnan, southern China, *Journal of SE Asian Earth Sciences* 9, 153-166 (1994).

91. Yin Jixiang. Current status and problems of investigation with reference to Gondwana stratigraphy in the Qinghai-Xizang Plateau and its adjoining areas. In: *Proceedings of the first symposium on the Qinghai-Xizang Plateau.* China Society of the Qinghai-Xizang Plateau Research (Ed.). pp.287-295. Science Press, Beijing (1992) (in Chinese).

92. Yu Guangming, Wang Chengshan and Zhang Shaonan. The characteristics of Jurassic sedimentary basin of Bangong Co-Dengqen fault belt in Xizang, *Bull. Chengdu Inst. Geol. M. R., Chinese Acad. Geol. Sci.* (13), 33-44 (1991) (in Chinese).

93. Zhang Qi, Li Dazhou and Zhang Kuiwu. Preliminary study on Tongchangjie ophiolite melange from Yun County, Yunnan Province, *Acta Petrologica Sinica* (1), 1-14 (1985) (in Chinese).

94. Zhang Qi, Li Dazhou and Zhang Kuiwu. Preliminary study of Paleo-Tethys ophiolites in Hengduan Mountain Region (HMR), China, *Journal of SE Asian Earth Sciences* 3, 249-254 (1989).

95. Zhang Qi, Zhang Kuiwu and Li Dazhou. *Mafic-ultramafic rocks in Hengduan Mountains region.* Science Press, Beijing (1992) (in Chinese).

96. Zhang Qi, Zhang Kuiwu, Li Dazhou and Wu Haiwei. A preliminary study of Shuanggou ophiolite in Xinping County, Yunnan Province, *Acta Petrologica Sinica* (4), 37-48 (1988) (in Chinese).

97. Zhang Qi, Zhou Yunsheng and Li Dazhou. Sheeted dike swarms within ophiolites in Xigaze-Baining district, Xizang, *Petrol. Res.* 1, 65-80 (1982) (in Chinese).

98. R. Zhang, B. Cong, S. Maruyama and J. G. Liu. Metamorphism and tectonic evolution of the Lancang paired metamorphic belts, southwestern China, *J. Metamorphic Geol.* 11, 605-619 (1993).

99. Zhang Yuquan, Xie Yingwen et al. *Geochemistry of granitoid rocks in Hengduan Mountains region.* Science Press, Beijing (1995) (in Chinese).

100. Zhang Zhimeng and Jin Meng. Two kinds of melange and their tectonic significance in Xiangcheng-Derong area, southwest Sichuan, *Scientia Geologica Sinica* (3), 205-214 (1979) (in Chinese).

101. Zhao Dasheng and Liu Xiangpin. Geochemical characteristics of collision type volcanic rocks from northwest Yunnan, *Geochimica* 23, 235-244 (1994) (in Chinese).

102. Zhao Jing, Zhong Dalai and Wang Yi. Metamorphism of Lancang metamorphic belt, the west Yunnan and its relation to deformation, *Acta Petrologica Sinica* (1), 27-40 (1994) (in Chinese).

103. Zhong Dalai et al. *Paleotethys: Lithosphere tectonic evolution in west Yunnan and Sichuan, China.* Science Press, Beijing (1997).

104. Zhong Dalai and Ding Lin. Gondwana dispersion and Asian accretion in view of Tethyan evolution in Southwest China and its adjacent regions. In: *Asian accretion.* Chinese Working Group of IGCP Project 321 (Ed.). pp.5-8. Seismological Press, Beijing (1993) (in Chinese).

105. Zhong Dalai, Ding Lin and He Fuxiang. The eastern Paleo-Tethys and its complicate continental margins. In: *Ann. Rep. Lab. Lithosphere Tectonic Evolution, Inst. Geol., Chinese Academy of Sciences (1993-1994).* Lab. LTE (Ed.). pp.7-11. Seismological Press, Beijing (1994).

106. Zhong Dalai, D. Tapponnier, Wu Haiwei et al. Large scale strike slip fault: the major structure of intracontinental deformation after collision, *Chinese Science Bulletin* 35, 304-309 (1990).

107. Zhou Weiquan and Lin Wenxin. Characteristics of blueschists in southern part of Lancangjiang metamorphic belt, *Regional Geology of China* (2), 76-85 (1982) (in Chinese).

108. Zhou Yiaoxiu, Lu Lianzhong, Yuan Xiangguo, V. Courtillot and J. Achache. Donnees paleomagnetiques nouvelles au Tibet-connaissance fondamentale pour le mouvement du Bloc de Lhasa. In: *Himalayan geology II.* Editorial Committee of "Himalayan Geology" (Ed.). pp.279-297. Geological Publishing House, Beijing (1984)

(in Chinese).

109. Zhu Zhiwen, Zhu Xiangyuan and Zhang Yiming. Paleomagnetic observation in Xizang and continental drift, *Acta Geophysica Sinica* **24**, 40-49 (1981) (in Chinese).

Proc. 30ᵗʰ Int'l. Geol. Congr., Vol. 6, pp. 77-86
Xiao Xuchang and Liu Hefu (Eds)
© VSP 1997

ARGUMENT ON RE-RECOGNITION ABOUT THE GEOTECTONICS OF SOUTH CHINA

ZHAO CHONGHE   HE KEZHAO   NIE ZETONG   LE CHANGSHUO
ZHOU ZHENGGUO   TAI DAOQIAN

(China University of Geosciences, Beijing, 100083)

YE NAN

(Geological Team of Northeastern Jiangxi, Shangrao, 334128)

Abstract

The geotectonic framework of south China has been one of the hotly discussed topics in Chinese geoscience since the 80's. Different views have been expressed regarding the tectonic evolution and forming age of the south China landmass, and among them whether there was a " Jiangnan old land " and the age of the " Banxi Group " were the focus of all arguments. Many scholars have put forward different viewpoints and tectonic evolution models from various angles, which may undoubtedly promote our understanding of the problem. Based on a series of new discoveries of recent years in NE Jiangxi and S Anhui, this paper attempts to clarify our viewpoints as follows:

(1)The "Banxi group" is not a pre-Sinian metamorphic stratigraphic unit.

(2)The "Jiangnan old land" did don't exist.

(3)A post-Caledonian platform didn't exist in south China; instead, there was a late Paleozoic ocean basin,which might form a part of the eastern Tethys.

(4)The Indosinian Movement left some important records in NE Jiangxi in terms of Late Paleozoic epimetamorphic rock and ophiolite complex.

Keyword: south China landmass, Orogenic belt, Ophiolite complex, Banxi group

1 THE FOCUS OF PROBLEMS AND ARGUMENTS

The age and tectonic evolution of the "Banxi group" are the major argument about the geotectonics of south China[1]. For a long time,most researchers have believed it to be belonging to Middle or Late Proterozoic, but there were also opposite viewpoints. All different viewpoints about the areas concerned are stated as follows,

1.1 The"Banxi group" was the pre-Cambrian foundation of the south China region

Ren Jishun et al. [2,3,4,5,6] believed that the"Jiangnan old land"was a big anticlinorium made up of polycyclic tectonics. Banxi group and Lengjiaxi group were not tectonic melange of uncertain age, but pre-Sinian strata (800Ma B.P) that had already been proved long before. In the Jiangnan anticlinorium and its nearby areas ultramafites should have resulted in pre-Sinian time,but definitely not oceanic basinal relics of Paleozoic-Triassic period. To this day, no Phanerozoic ophiolite complex has been found from within the areas concerned, therefore, there is no orogenic tectonic melange belt.

Guo Lingzhi et al.[7,8,9,11], and Xu Bei et al.[10] thought that the late Proterozoic palaeogeographic framework of south China was an arc-basin system, which consisted of the SW Anhui terrain, the NE Jiangxi intermediate block and the Zhe-Gan island-arc terrain. SW Anhui represented the late Proterozoic continental margin extending to the south, with back-arc deposits developed to the north and the Anhui-Zhejiang-Jiangxi island-arc terrain to the south of the marginal sea. In middle or late Proterozoic the major tectonic background of the Anhui-Zhejiang-Jiangxi region was that the Cathaysia plate slopingly underthrust the Yangtze plate, the subduction zone truncated the southeast continental margin of the Yangtze plate, and that the region to the south of island-arc terrain might be the front of the subduction zone of the the Cathaysia plate. This region had been subjected to two collision events(510Ma;429Ma) in the Caledonian period, so that the deformation system of this orogenic belt shows two reversely thrusting-"dual-thrust collision orogenic belt" system.

Wang Hongzheng et al.[12,13] pointed out that in late Proterozoic and Sinian the region was characterized by the Banxi group, which was turbidity current deposits of passsive continental margin,and in Jingninian age it was underthrusting to the north. The existence of a Paleazoic ocean was inconsistent with the facts. In post Caledonian the south China region had become part of the mature continental crust, large scale collisions and nappes were all of intra-continental nature during the Mesozoic era .

1.2 The"Banxi group" is the relict of nappe structure in the Mesozoic orogenic belt

J. Kenneth Hsu et al.[14,15,16,17,18] consider that south China wasn't a post-Caledonian platform , but a composite orogenic belt, which was mainly formed after the Indosinian orogeny. The Nanpanjiang ocean had appeared between the Yangtze-and south China block, which had deep-sea deposition. The Banxi group, instead of being strata, is a relict of the eroded nappe which was thrust from the southeast upon the carbonatite platform. It was tectono-melange and was characterzed in some places by ophiolite melange.

The present authors have also noticed the recent researches of Li Rijun et al. [19] and Chen Xu et al.[20], their viewpoints belong to the former (1.1).

Evidently the above-mentioned opinions can be divided into two kinds of basic viewpoints,although they differ in principle regarding geologic bases, division of tectonic system and the model of tectonic evolution. The focus and essence of argument lie in the knowledge of the age and tectonic meaning of the "Banxi group" in south China. The premise of the former opinion is to regard the Banxi group as a pre-Sinian metamorphic stratigraphic unit, while the latter is to take it as a kind of tectonic melange. These are the two completely opposite viewpoints on the knowledge of south China tectonics. Therefore, in the light of recent years' new achievements, it would be of great and far-reaching significance to deepen the discussion of the problems of south China tectonics. Based upon the present state of research on the "Banxi group", there are still some problems to be solved: ① The age of the original "Banxi group" is based mainly on discordance and metamorphic grade. Although later studies discovered fairly rich paleo-microplants and isotopic datings, there are questionable unconformities along the discordance of the "Banxi group", which extends over five provinces for about 1500Km long. So,the age and correlation of parts of this group--pre-Sinian or not--have to be carefully studied further. ② As already noticed by some scholars the Middle and Late Proterozoic strata of south China are dispersely distributed; moreover, they differ greatly in such aspects as lithology, lithofacies and tectonic background. The original "Banxi group" has been partially disintegrated and has been named with different names for their distinction.Because of the lack of re-recognized age evidence (e. g. lacking the generally accepted index fossils), it is difficult to make regional correlation. ③ In recent years, Paleozoic fossils have been discovered one after another in the epimetamorphic rocks in S Anhui and NE Jiangxi, (eastern of south China), where the Banxi group has been considered for a long time to be belonging to the Proterozoic. Hence the Banxi group is not a metamorphic stratigraphic unit of the Proterozoic. However, further study is needed to decide whether the age of this suite of epimetamorphic rocks should be changed or this suite of epimetamorphic rocks contains the strata of different ages.

2 PROGRESS AND DISCOVERIES

In recent years, Paleozoic fossils have been discovered in the so-called Middle or Late Proterozoic epimetamorphic strata in S Anhui and NE Jiangxi(Fig.1). These new and important findings provide new evidence to promote our knowledge of the geotectonic evolution of south China.

Xu Shutong et al.[21] discovered some fossils in the Proterozoic epimetamorphosed

80

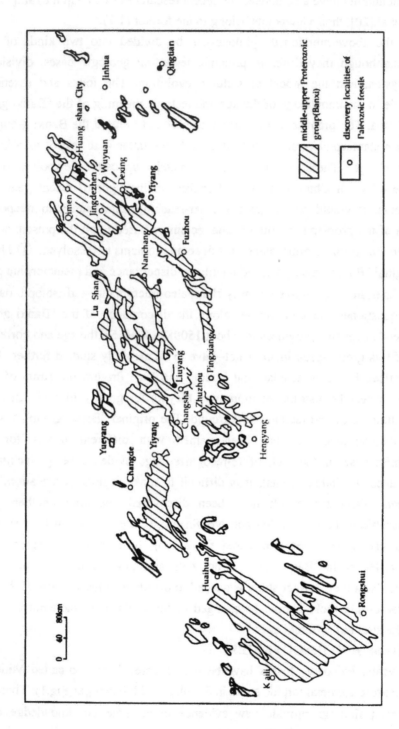

Fig. 1 Distribution of the "Banxi group" and the discovery spots of middle Paleozoic fossils in south China.

Shangxi Group in Xiuning and Qianxian, S Anhui. One of the sites is in Xiadanian and Zhangqian-Shangxikou of Xiuning county, where *Baltisphaeridium* sp.(algae)and *Lingulacea* (brachiopod fam. et gen.indet.) of early Paleozoic were discovered from slates of the originally named Shangxi group; the other site lies near Anxi, Qianxian and Feishuxia, Yangzhanling on the northern margin of the original "Jiangnan old land",where spores and Conularid(?) were discovered in Xiuning sandstone and Niuwu formation of the original Sinian group. The fossils should belong to late Paleozoic.

Zhao Chonghe, He Kezhao et al.[22,23,24] had scored a breakthrough in the study of the NE Jiangxi faulted belt and its neighboring areas since 1993. Firstly in 1993, they found for the first time late Paleozoic radiolarian-bearing silicolite in two sites of the epimetamorphic ophiolite suite(formerly considered to be middle-late Proterozoic strata) in the Zhangshudun mine of Yiyang.Though the rock had become blastic, the shape of radiolarians are intact, which are of Carboniferous-Permian in age according to Wu Haoruo et al. (1995). In 1996, based on the above discovery, He Kezhao et al. found exposures of late Paleozoic radiolarian silicolite at three to four sites over a larger scope in NE Jiangxi-fault zone. It is especially significant that phyllite interbedded with late Paleozoic radiolarian-bearing silicolite are included in the original Sinian system, located in Dengshan village outsides NE Jiangxi fault-zone. Secondly, in Zhangshudun and Yejiachun in NE Jiangxi. Recently dating data of 413.16Ma and 419.32Ma by means of $^{40}Ar/^{39}Ar$(dated by Dai Tongmo,1996) from the basic volcanic rocks,which were regarded for a long time as middle-late Proterozoc. Besides, Zhu Xianjia got the datings of 398, 380, 375, 292, 255Ma(1986) by the K-Ar method from the basic volcanic rocks in the same area. Thirdly, the radiolarian-silicolite recently found is intercalated in phyllite and this suite contains neither terrigenous fragments nor carbonate rocks, and is typically deep water silico-argillaceous sediments.Its petrofacies, geochemical and $\delta^{30}Si$ isotope all reflect that it was a product of tectonics and palaeogeographic environment and is different from the so- called platform silicolite (the Permian Gufeng and Dalong formation).

Xue Chongsheng et al.[26] found microfossils in the ophiolitic melange and epimetamorphic rocks of the Zhangchun Dengshan groups in NE Jiangxi which marks also a significiant progress. They found successively Sponge-spicule, Worms and Archaeoides in the rocks in Zhujia, Qigongzheng of Yiyang,and found vase-like species, operculates, *Archaeoides* and *Acritarchs* fossils in tuffaceous fine siltstone in Zhangshudun and found similar Archaeoides in tuffaceous silt sandstone in Sipu, Shangkuqian and Huangtuling of Hengfeng county. Xue Chongsheng et al.[26] considered that similar blocks of Zhangshuchun -Dengshan group and Zhangshudun ophiolitic melange and the flysch formation with baume sequence in Sipu, Shangkuqian, Huangduling are products of the same age, which is early Paleozoic.

In addition to the areas mentioned above, Paleozoic fossils were also found in NE Jiangxi and SE Zhejiang and E Hunan where strata were formerly considered to be Proterzoic and Sinian. Di Ruiji et al.[25] found fossil sponge spicules in late Proterzoic silicolite of spilite intercalation in the Huangma-Baitu area of Fengcheng county , and considered it to be early Paleozoic. In 1993, the Second Team of Coal Geology of Zhejiang found Baridio and fossil fish bones in marble which was originally regarded as Proterzoic Hexi group in Zhixitou, Qintian county and considered it as Permian. Xu Butai et al.(1989) got a dating of 231 ± 7Ma of whole rock isochronism, and took it as Permian-Trias. Also in Liuyang, E Hunan provience, Liu Zuzhi found *Frondicularia* Defrance and algal fossils in strata which were originally considered to be Sinian in the Yonghe phosphate mine district of Liuyang in E Hunan ,and he thought it to be late Paleozoic(private communication,1996).

The new discoveries mentioned above only mean a beginning of the research on the Proterozoic group and the Sinian system in the south China region. They are neither accidental nor isolated, but the inevitable results of the ever deepening process of geologic research on the region. Therefore, to this day, the following conclusions may be given: ① the Proterozoic group of south China, especially the original"Banxi group" at least a part of them, does not belong to the Proterozoic group but rather to the Paleozoic group; ② The "Jiangnan old land" didn't exist. The oceanic crust disappeared after but not before the Paleozoic; ③ Parts of strata in Anhui-Zhejiang, Jiangxi, Fujian of the south China region, which were originally regarded as Proterozoic, are subjected to change in age, because of the discoveries of Paleozoic fossils in them; besides, a considerable part of sedimentary environments doesn't belong to the stable deposits of platform type, but is characterized by strata of orogenic belt. With the development of theory, research method and analytical technology in paleontology and stratigraphy, it is imperative to disintegrate the "Banxi group" extending for 1600 killometres long and further deepen the knowledge of its nature.

3 RE-RECOGNITION ABOUT THE GEOTECTONICS OF SOUTH CHINA

The new discoveries from the original "Banxi group" in Anhui, Jiangxi, Zhejiang and Fujian provinces of south China have doubtlessly shaken some traditional opinions and theories. Although the above- mentioned discoveries are as yet insuffcient for putting forward a systematic opinion and model for the geotectonics of south China, they have provided us a new train of thought to re-recognize the geotectonics of this region, and also the evidence of geologic events ignored. To sum up, several points are given as follows:

3.1 The "Banxi group" is not a pre-Sinian metamorphic stratigraphic unit.

This view can be proved by the above-mentioned new facts. In addition,some scholars also quarreled from different angles. Shu Liangshu et al. [11] believed that the nearby Jiuling and Huaiyu block respectively have the unique history of geotectonic evolution and stratigraphically, there are not connected by transitional beds. The differences are even greater in some areas far away, and the explanation to their cause as we have investigated in the field is that these epimetamorphic rocks occur in fact as lots of larger or smaller tectonic blocks, the stratigraphic sequence of which is continuous in some local areas, but generally uncontinuous. In the regional geological surveying in east Shangrao, Li Changnian et al. [27] also found that the so-called Proterozoic strata are tectonic blocks. So, at least, it can be said that the Banxi group in relatively larger area in the eastern part of south China is a suite of orogenic stratigraphic blocks, whose sequence should be studied by non-Smithian stratigraphic methods. The age of these strata has been discussed in foregoing paragraphs, but more works shoud be done further.

3.2 The Jiangnan old land didn't exist.

As the age of the strata in south China,which was originally considered to be Proterozoic, is questionable and at the same time, these strata are orogenic stratigraphic blocks, the Jiangnan old land used to be so called, didn't exist, at least in S Anhui, N Jiangxi, W Zhejiang and NE Fujian, the eastern part of the south China region, and its evolution should be reconsidered.

3.3 The Post-Caledonian platform didn't exist.

Was there deep-sea sediment in late Paleozoic in south China? Was there any ocean basin? These are key problems related to the geotectonic evolution of south China. The discovery of late Paleozoic radiolarian silicolite in S Anhui and NE Jiangxi provinces, as discussed in the foregoing paragraphs, provides new and important evidence for solving these questions. According to comprehensive analysis of rock association, geochemistry and isotope $\delta^{30}Si$ ‰ of the ophiolite suite, the silicolite containing radiolarians as discovered in Yiyang, it can be concluded that the rock association, which has undergone epimetamorphism, is completely different from that of the Carboniferous-Permian strata in the middle and the lower reaches of the Yangtze River, which is definitely not the persistent sediment in a platform, but the deep-sea sediment in an ocean basin, and probably a part of the Paleotethys.Whether it was the Nanpanjiang ocean called by J.Kenneth Hsu et al. [17], or the intracontinental aulogen called by Ma Wenpu [28], needs to be worked out further. But it can be confirmed that the existence of the late Paleozoic deep-sea sediments is beyond doubt [22,23,24]. Owing to the convergence of the plates, the ocean basin ever existing in the late Paleozoic disappeared, its trace has not always been noticed, and its existence has been neglected. This is the main reason why recently there have been arguments about the geotectonics of south China. However, the new evidence has proved that the extent of the late Paleozoic ocean basin of south China definitely covers an area of certain extent.

3.4 The Indosinian movement established the tectonic framework of south China.

Based on the characters of the outcrops with the late Paleozoic radiolarian silicolite and phyllite, the epimetamorphism of silico-pelitic sedimental formation was a regional geological phenomenon, which shows that a large geological event took place after late Paleozoic, that was, the Indosinian movement had played an important part in this area. The regional metamorphic rocks of late Paleozoic deep-sea sediments and the ophiolite melange distributed along the NE Jianxi fault-belt are its rock records.

3.5 It is of practical significance to recognize again the geotectonic framework of south China. The new geological discoveries offer us the opportunity to carry on further research on the geological tectonics of south China, and to re-establish the geotectonic framework of this region. The importance of such a study is concerned not only with theory, but also of strategic practice in mineralization, especially in the deployment of strategic searching for hydrocarbon and exploring solid mineral resources.

Thanks are due to Professor-Academician Yang Zunyi for his advise. The research project is funded by National Natural Sciences Foundation of China.

REFERENCE

1. Jiangxi Bureau of Geology. *Regional Geology of Jiangxi Province*, pp. 853 Geol. Publ. House, Beijing. (1985).

2. Ren Jishun. *The Geotectonic Evolution of China*: Explanatory Note for 1:4,000,000 Tectonic Map of China. Sei. pp. 124. Pub. House, Beijing (1980).

3. Ren Jishun. Chen Tingyu and Liu Zhigang. Some Problems on the division of tectonic units in eastern China, *Geol. Review*, 30:4, 382-385 (1984).

4. Ren Jishun. Chen Tingyu and Liu Zhigang. On geotectonic problems of South China. *Chinese Science Bulletin*, 1, 49-51(1986).

5. Ren Jishun. On the geotectonics of South China. *Acta Geol. Sinica*, 64:4, 275-288(1990).

6. Ren Jishun, Chen Tingyu and Niu Baogui et.al.. *Tectonic evolution of the continental lithosphere and metallogeny in eastern China and adjacent areas*. pp. 205. Sei. Publ. House, Beijing (1992).

7. Guo Lingzhi, Shi Yangshen and Ma Ruishi. *The tectonic framwork and crustal evolution of*

South China. Scientific Paper on Geology for International Exchange, Prepared for the 27th International Geological Congress. (4). pp.109-116. Beijing; Geol. Publ. House (1980).

8. Guo Lingzhi, Shi Yangshen and Ma Ruishi. *The study of terrane tectonics in southeast China.* Jour. of Nanjing University (Nature Scince), 20:4, 732-737(1986).

9. Guo Lingzhi, Shi Yangshen and Ma Ruishi et. al.. *The Plate movment and crust evolution of the Jiangxi Proterozoic island-arc tectonics.* Proceeding of International Symposium on Precambrian Crustal evolution No. 1: Tectonic. pp30-39. Beijing: Geol. Publ. House (1986).

10. Xu Bei, Guo Lingzhi and Shi Yangshen.*Proterozoic terranes and multiphase collision orogens in Anhui-Zhejiang-Jiangxi area.* pp. 112. Beijing: Geol. Publ. House (1992).

11. Shu Liangshu, Shi Yangshen, Guo Lingzhi et. al.. *The late Proterozoic plate tectonics and Collisional Kinematics in the middle part of the Jiangnan belt, South China.* pp. 174. Nanjing University Press (1995).

12. Wang Hongzhen. *A synopsis of the tectonic development of South China*(Editors: Wang Hongzhen et.al.) pp. 1-15. Wuhan College of Geology Press (1986).

13. Wang Hongzhen, Yan Sennan, Liu Benpei et.al.. *Tectonopalaeogeography and Palaeobiogeography of China and adjacant regions.* pp. 347. China University of Geoscince Press (1994).

14. Hsu K.J. Thin-Skinned Plate-tectonic model for Collision type orogensis. *Sci. Sin.*, **24, 100**, (1981).

15. Hsu K.J., Sun Shu and Li Jigliang. Huanan Alps, not South China Plateform. *Sci. Sin.,* **Ser.B(1987)**, 1107-1115 (in Chinese)(1987).

16. Hsu K.J.. The problems of geotectonic on South China. *Geol. Sci. and Technology Information*, **62**, 13-27(1987).

17. Hsu K.J., Li Jiliang ,Chen Haihong et.al.. Tectonics of South China: Key to understanding West Pacific geology. *Tectonophysics,***183**, 99-39 (1990).

18. Li Jiliang (Ed.). *Lithospheric and tectonic evolution of Southeast China.* pp. 264. Beijing: Metallargical Industry Publ. House (1993).

19. Li Yuejun, Hao Jie and Lu Gangyi. The Banxi group and the Banxi ophiolite melange. *Geol. Review,* **40:2**, 97-105(1994).

20. Chen Xu, Rong Jiayu and D.E. Rowley et.al.. Is the early Paleozoic Banxi ocean in South China necessary?. *Geol. Review,* **41:5**, 389-400(1995).

21. Xu Shutong, Chen Guanbao, Tao Zheng et. al.. The fossils in Shangxi group and its implication for tectonics, soutern Anhui, China. *Scince in China,* **Ser. B, 37:3**, 366-373(1994).

22. Zhao Chonghe, He Kezhao, Mo Xuanxue et. al.. Discovery and its significance of late Paleozoic radiolarian silicalite in ophiolitic Melange of Northeastern Jianxi deep fault belt. *Chinese Scince Bulletin,* **41:8**, 667-670 (1995).

23. Zhao Chonghe, He Kezhao, Mo Xuanxue et .al.. *The characteristies of the Northeastern Jianxi ophiolitic melange belt and its tectonic implication. In: Study on ophiolites and geodynamics* (Editor by Zhang Qi). pp. 208-212. Geol. Publ. House, Beijing, China (1996).

24. He Kezhao, Zhao Chonghe, Tai Daoqian et. al.. Discovery of the late palaeozoic radiolarian silicolite in many places in Northeastern Jiangxi ophiolitic melange belt. *Geoscinnce*, 10:3, 303-360 (1996).

25. Di Ruijie, Shen Jun and Jiang Jinyuan et.al.. Discovery of Sinian-Early Cambrian strata near Nangchang, Jiangxi Province, *Geol. Reviw*, 35:6, 537-542(1989).

26. Xue Chongshen, Zhang Kexin and Zen Zongping et. al.. Discovery microfossil on ophiolitic melange and metamorphic strata of Zhangcun group-Dengshen group, Northeastern Jiangxi Province. *Geological scince and Technology Information*, 15:1, 30 (1996).

27. Li Changnian, Xue Ghongsheng, Xhang Kexin et.al.. The ophiolites in northeastern Jiangxi province and their geodynamical significance. *Study on Ophiolites and Geodynamics*(Ed. by Zhang Qi). pp. 213-217. Beijing: Geol. Publ. House (1996).

28. Ma Wenpu. Paleothethys in South China, Permian orogeny and the eastwards extension of interchange domain, *Scientia Geol. Sinica*. 31:2, 105-113(1996).

Proc. 30ᵗʰ Int'l. Geol. Congr., Vol. 6, pp. 87-104
Xiao Xuchang and Liu Hefu (Eds)
© VSP 1997

Evolution and Structural Styles of the Sichuan Foreland Basin

Cai Liguo
Institute of geology. Chinese Academic of sciences Beijing 100029 China
Liu Hefu
China University of Geosciences Beijing 100083 China

Abstract

The formation of the Sichuan foreland basin was the result of rifting and accretion in the border and interior of Yangtze block. The development of the foreland basin can be divided into two stages: (1) the early stage (O_2-S), the formation of collision-type foreland basins; (2) the late stage (T_3-Kz), the development of subduction-intracontinent-type foreland basins. These different types of foreland basin were superimposed and formed a composite foreland basin. The collision-type foreland basin was formed by subduction and closure of oceanic crust, and accompanied with ophiolite, melange and arc. Flysch was deposited in the early and molasse deposited in the late, and composed of the two-layer filling structure. The intracontinental subduction-type foreland basin was not developed in the convergent plate boundary but in intraplatform, not accompanied with ophiolite, melange and arc, and related with intracontinental subduction. Molasse was deposited in the basin and formed one-layer filling structure.
The main structural styles were controlled by the thrust and fold belts in the foreland basin. The structural deformation clearly showed the plane and vertical zonation. The main structural styles were developed in the foreland basin. They are thrust-nappe structure of basement, imbricated thrust system, nappe structure, duplex, pop up, triangular zone, detachment fold, fault-bend fold and fault-propagation fold.

Key words Sichuan foerland basin, Yangtze block, thrust-fold belt, structural style

The evolution of the Sichuan foreland basin

The formation of the Sichuan foreland basin was the result of rifting and accretion taking place in the border of Yangtze block. The basin was controlled by the thrusting of Longmenshan, Dabashan, Chuan-Xiang-Qian and Kang-Dian thrust and fold belts. Therefore, the basin is a tectonic basin and about $18,000km^2$ in area. The development of the basin can be divided into four stages as follows: (1) the development of passive margin basin in the early Palaeozoic age, (2) the formation of foreland basin related to collision orogeny in the latter of early Palaeozoic age, (3) the evolution of intracontinent rifting basins in the late Palaeozoic, and (4) the development of foreland basin related to intracontinent subduction. Therefore, the basin is a tectonic basin superimposed by different proto type basins.

The breaking up of Yangtze block and formation of passive margin basins($Z-O_1$)

The continental rifting occurred in interior and margin of Yangtze block, with continental detrital rocks, turbidites and volcanic turbidites, in the late Proterozoic age, and showed the breaking up of Yangtze block. And then Yangtze block had been developed into three different passive margins in its periphery: (1) the passive margin of Qinling-Dabie; (2) the passive margin of southeastern Yangtze; and (3) the rifting margin of western Sichuan and Yunnan(Chuanxi-Dianxi) during the late Sinian period to the early Ordovician age (Fig. 1).

There had been three tectonic-sediment lithofacies on southeastern Yangtze passive margin from Sinian period to the early Ordovician period: (1) abyssal-hemipelagic black shale, carbargillite and carbon-bearing silicalite, about 2000–3000 meter in thickness at Jiangnan-Xuefeng area; (2) turbidity deep depression, south between Guangfeng-Pingxiang and Jiangshao faults, and was filled with about 20,000 meter distal turbidite interbedded black silicalite, contourite, volcanic rocks in local; and (3) the melange belt, formed by oceanic crust blocks, continental crust blocks and deep sea flysch(Wang Xin, 1988), distributed along the Lishui-Zhenghe-Dapu fault zone and presented the existing of oceanic crust or transitional oceanic crust in that time. The melange presented the boundary between Yangtze block and Huaxia continental block. There would be several micro-continental blocks distributed on the passive margin, and turbidite was deposited (Qin Deyu, 1993). This showed that the passive margin was irregular.

The Qinling-Dabie passive margin which was consisted of uplifts and rifting basins and was filled with about 1500–4000 meter of abyssal black silicalite, carbon-bearing silicalites, carbonaceous shale. The shallow marine carbonates interbedded sandstone and shale was deposited on the uplift. The geochemical analysis shows that the ophiolites, outcropped along Danfeng-Xinyang suture zone were formed in typical mid-oceanic ridge. The isotopic age of basic volcanic rock is about 560±61Ma and 570Ma (Qin Deyu, 1988). The ophiolite pebble is found in the Devonian conglomerate (Zhang Guowei, 1988). These show that the oceanic crust was formed in the early Palaeozoic age. Therefore, Danfeng-Xinyang suture zone was the boundary between Yangtze block and Huabei block.

The Chuanxi-Dianxi rifting margin between Luhuo-Daofu-Ailaoshan fault zones and Lancangjiang-Shuangjiang suture zone had formed from the late Proterozoic to Carboniferous period. The platform carbonate rock was distributed from east through Ganzi-Litang fault zone to west. A group of metavolcanite, beneath the Sinian period algal dolomite, has been found in the west of Ganzi-Litang, and similar to the Suxiong Formation in western Sichuan. This shows that the rifting activity began in the area in the late Proterozoic age. During the Cambrian-Ordovician period, the rifting passive margin had declined to west, formed block-uplift and rifting basin, and were deposited mudstone, microclastic rocks, argillaceous limestone and limestone of shallow marine to deep sea basinic facies. The volcanic rocks, belonged to intracontinental rifting volcanic eruption (Chen Bingwei, 1987), were found in Jinshajiang and Ganzi-Litang rifting basins in the lower Cambrian and the lower Ordovician sediments. The biota of the early Palaeozoic age(Cambrian and Ordovician) in Changdu block was similar to that found in Yangtze block. This shows that Changdu block would be a part of Yangtze block at this time.

In the Palaeozoic age, an open shallow marine sedimentary environment occupied the Yangtze platform that was surrounded by passive margins. The Carbonate rocks of platform facies, included dolomite, gypsum and halilyte, tidal-flat clastic rocks, limestone, open platform facies limestone and marginal organic reef and oolitic bank were deposited. The largest subsidence event emerged in Yangtze platform in the early Cambrian age, and

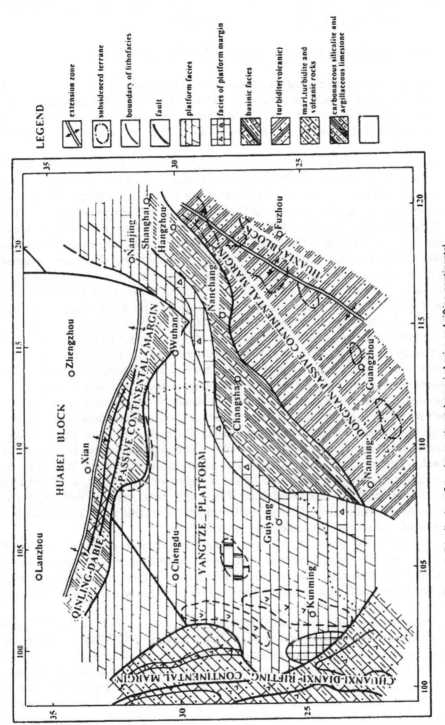

Figure 1. Distribution of passive continental margin basins and rifting continental margin in Yangtze Block during Sinian-middle Ordovician period

black carbargillite, cabonaceous-pelitic silicalite and phosphate rock accumulated on the platform, which formed a favorable source rock in the basin.

The formation of foreland basin in the late Paleozoic period(O_2–S)

The Qinling-Dabie oceanic crust had been subducted towards north under the Huabei block, and Dongnan oceanic or transitional oceanic crust subducted towards southeast under the Huaxia block respectively from the late Ordovician to Devonian. The foreland basins associated with collision were formed on the margins of northern and southeastern Yangtze block (Fig. 2).

The subduction and collision of oceanic or transitional oceanic crust of Huanan and micro-blocks markedly began in the middle-late Ordovician, and the passive margin was reversed. The sediment of the upper Ordovician and Silurian was absented along the east of Jiang-Shao fault zone. This presented the influence of collision between Huaxia block and Yangtze block. A foreland flysch formed in the late Ordovician age in western Zhejing (Yuqian Formation). The littoral delta, foreland flysch, deep basin of radiolarian silicalite, carbargillite was deposited from southeast to northwest. The clastic sediments in the deep water turbidite were well graded, but poorly rounded, lower maturity and the rate of sedimentation was 250m/Ma (Zhang Fuxiang, 1993). Then the turbidite was migrated westwards from southern Jiangxi to Shuangfeng west of Hengyang with time, at the end, the foreland flysch turbidite was migrated northwards to Xupu, Xinhua in the central Hunan Province.   This showed that Yangtze continental block was being influenced by the collision of Huaxia continental block, and the depocenter and center of subsidence were continually migrated towards northwest. The supply of bilateral clastic sediments showed not only the existence of Huaxia block but also a forebuldge in southeastern Yangtze block, and the forebuldge would be the central Guizhou uplift. In the end of Silurian, except relict oceanic basin in Qinzhou-Fangcheng area, the ocean between Huaxia and Yangtze blocks was closed. The early Palaeozoic passive margins were revered and superimposed a foreland basin. The sediments of passive margin were involved in deformation and denudation. The sediments of middle and lower Devonian were absented in Jiangnan uplift and Lower Yangtze area. However, an alluvial And black silicalite mudstone interbeded some fine-clastic rock, volcanic rock, terrigenous turbidite were deposited in the Qinzhou-Fangcheng relict oceanic basin. This would be an indication of micro-rifting in southern margin of Yangtze.

The collision taking place in southern margin of Yangtze block resulted in the deformation of Caledonian recumbent folds and imbricated thrusts in Jingnan-Xuefeng area. Under the influenced of the orogeny a large scale of regression occurred in Yangtze block, and the front of thrust and fold belt would be located in western fringe of Jiangnan uplift.

The Qinling-Dabie oceanic crust subducted along the suture zone of Danfeng-Shangnan-Tongbai-Xinyang in the late Ordovician age (Gao Changlin, 1993). The early Palaeozoic passive margin was changed into an active margin with trench-arc system in southern Huabei block. A series of volcanic rocks was formed in the island-arc (Danfeng Group) and back-arc basin (Erlangping Group).After strong subsidence taking placed in the late Ordovician, the early Silurian age the passive margin was filled up. A large thickness of detrital rocks, mudstone interbeded carbonate rocks were quickly deposited above the lower part of the lower Silurian deep water turbidite, carbonaceous silicalite, sandyshale.

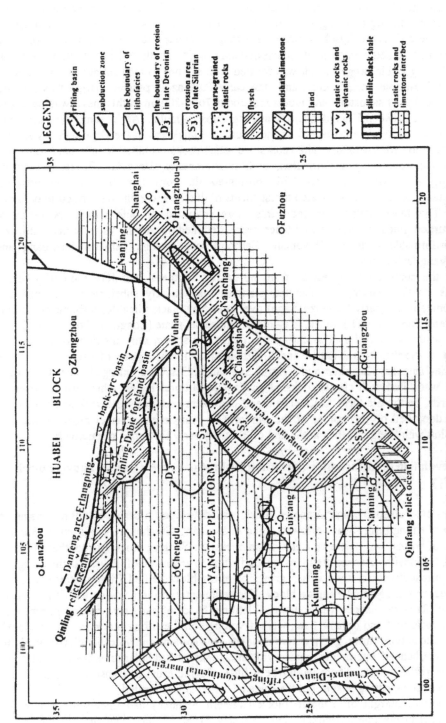

Figure 2. Distribution of foreland basins and rifting continental margin in Yangtze Block during middle Ordovician -Silurian period

LEGEND

- rifting basin
- subduction zone
- the boundary of lithofacies
- D_3 the boundary of erosion in late Devonian
- S_3 erosion area of late Silurian
- coarse-grained clastic rocks
- flysch
- sandshale,limestone
- land
- clastic rocks and volcanic rocks
- silicralite,black shale
- clastic rocks and limestone interbed

The load mold, flue cast, parallel bedding, little cross-bedding and lamination developed in western Qinling (Cai Liguo, et al. ,1993). These all showed that the Silurian system would be a foreland flysch, which was strongly contrary to the early Palaeozoic non-compensate of sedimentation in the passive margin. In the end of Silurian-early Devonian the Qinling-Dabie orogenic belt was formed with the northwards subduction and closure of Qinling-Dabie oceanic crust and the early Palaeozoic passive margin was revered. The turbidite of sandstone and mudstone, about 7000-8000m in thickness, was deposited on the area between Shangdan and Shanyang-Zuoshui faults. At the same time carbonate and clastic rocks were deposited in the South, the thickness of the Devonian system was reduced to 3000-4500m. The coarse clastic rocks composed of sandstone and conglomerate (Tongyu Formation, at Zhouzhi county; Dacaotan Group, in western Qinling area) was deposited. This presented the formation of collision orogenic mountains. A series of imbricated thrust sheets of ductile shearing was formed in orogenic belt. The lower Palaeozoic sediments were involved in deformation. A series of overturned folds, whose axial planes were declined to north, was formed and was uncomformable with the Devonian- Carboniferous system, between Shandang and Shanyang fault zones.

With the continual compression from north and southeast and formation of foreland basins on the margin, the Yangtze platform was quickly subsidence. The platform carbonate rocks (early Palaeozoic age) overlapped by black silicalite, carbonaceous shale (Wufeng Formation, O_3), black, gray-black sandshale, shale (Longmaxi Formation, S_1) of deep water basin facies. The carbonate rocks of platform only distributed in southern Sichuan, Northeastern Yunnan and Guizhou area. In the upper of early Silurian the Huaxia-Jiangnan, Kangdian and Central Sichuan uplifts were enlarged. The deep water basin was filled with clastic rocks of shallow sea-littoral facies, which was coarsening upwards, on the platform. Only clastic and carbonate rocks of shallow sea-littoral facies were deposited on the border of the platform as Huaxia and Huabei blocks came into collision with Yangtze block at the end of Silurian-early Devonian respectively.

The evolution of intracontinental rifting basin with breaking up of Yangtze block ($D-T_2$)

In the late Palaeozoic age ,Yangtze block was developed a series of rifting basins. A group of rifting basins was northwest trend in Yunnan, Guizhou, Guangxi area. Another group of basins was northeast trend in southeastern Yangtze block and southern Qinling rifting basin with east-west trend, associated with the extension of Palaeo-Tethys (Fig. 3). The typical structural style was a group of micro-blocks rifted from southwestern and southeastern margins of Yangtze block. The sedimentary lithofacies on the rifting basins and uplifts were changed very quickly. Platform facies of carbonate rocks, interbedded clastic rock, reef and organic bank were developed on the rifting block (or micro-block). Carbonate gravity flows, gravity-slide deposits, terrigenous and volcanic clastic turbidite, often interbedded intermediate-basic volcanic rocks and radiolarian silicate, were filled on the rifting basins. There would be a little oceanic crust or transitional crust in the margin of some rifting basins, but the scale of the crust was limited.

The western margin of Yangtze block (western Sichuan) had undergone two major extensions in the late Palaeozoic and the early Mesozoic age. The first extension had taken place in the Carboniferous-early Permian age, the result was rifting of the Changdu

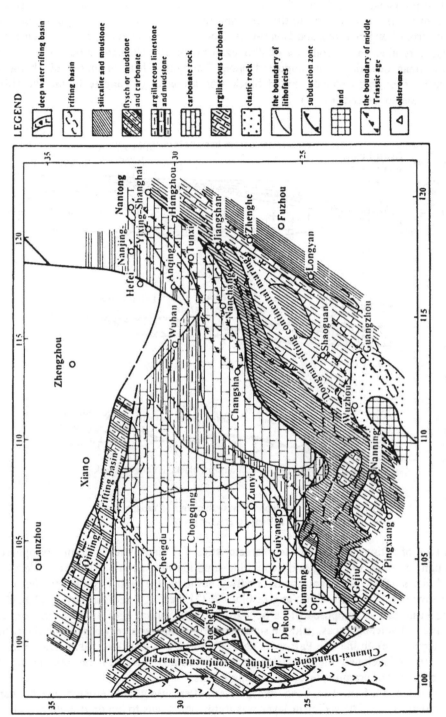

Figure 3. Distribution of rifting basins in Yangtze Block during Devonian-middle Triassic period

LEGEND

- deep water rifting basin
- rifting basin
- silicalite and mudstone
- flysch or mudstone and carbonate
- argillaceous limestone and mudstone
- carbonate rock
- argillaceous carbonate
- clastic rock
- the boundary of lithofacies
- subduction zone
- land
- the boundary of middle Triassic age
- olistrome

block from Yangtze block and the formation of Jinshajiang oceanic basin. The second occurred in the early Permian to the early and middle Triassic age, Zhongzan block was broken from Yangtze block and Ganzi-Litang oceanic basin developed. The extension of the latter resulted in the collision of Zhongzan block with Changdu block. Then, as the Ganzi-Litang oceanic basin subducted under Zhongzan-Changdu united block (at the Norian), Zhongzan-Changdu united block was accreted to Yangtze block, the rifting margin came to the end (Fig. 4).

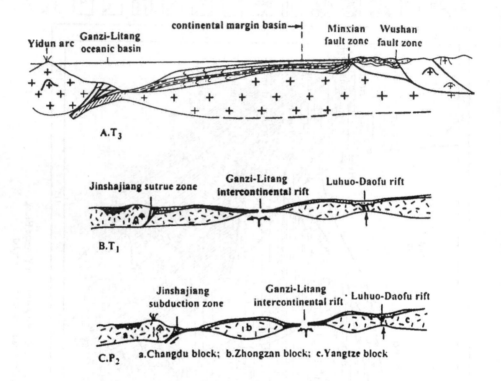

Figure 4. The tectonic evolution of western margin of Yangtze Block during the Late Palaeozoic-Triassic period

This type of rifting margin has the characteristics: (1) there is not a complete ophiolite although sea-floor spreading and submarine eruption may occur; (2) there is deep sea radiolarian silicalite on oceanic basin or rifting basin, but the biota is similar to each other and shows a relationship with Yangtze block's; (3) there is a sedimentary "melange" in rifting basin that is composed of carbonate blocks of variant ages mixed with deep sea turbidite and volcanic rocks. ,and the sedimentary melange was very popular in the rifting of Yangtze block in the late Paleozoic age; (4) trench-arc -back-arc basin and volcanic arc were undeveloped along the main suture zones, this may be related to the undevelopment of oceanic crust in the rifting basins.

The Yangtze platform had been peneplained after lifting and erosion in the early Carboniferous age. The sediment layer was think, about tens to hundred meters, and was stable distributed stably in middle and lower Yangtze area, most of them were platform carbonate rocks, and a few belonged to continental fluviolacustrine facies. Under the influence of intracontinental rifting some rifts were developed inside the platform in the latter of early Permian and the late Permian age. The deep water radiolarian silicalite and argillaceous silicalite were deposited on the rifts, and would be tracked from Nanjing (Jiangsu), Chaoxian, Jingxian (Anhui), Jingshan(Hubei) to northern Sichuan. The platform organic reef-banks and carbonates rocks were developed on the Yangtze platform. The platform was deeper northward as the subsidence, and deposited from carbonate rocks on Jiangnan platform, oolitic bank on platform margin to shelf-slope brecciola and shelf argillite in the early Triassic age. The platform had been shriveled since the middle Triassic age, and evaporite of tidal flat-lagoon facies and continental marine facies changed generally to continental clastic rocks. It is clearly that the terrain suture zones of southeastern margin on Yangtze block had been formed.

The uplift of central Sichuan and Jiangnan had not been deposited during the Devonian to early Carboniferous age, until the late Carboniferous period the two uplifts were deposited platform dolomite, limestone with the sea level rising. Then, under the influence of rifting, the carbonate platform was broken up along the northeast and northwest trends of faults. The rifting uplifts and rifting basins were formed in the margins in the latter of early Permian and the early of late Permian age. Carbonate rock was developed on the rifting uplifts and deep water silicalite, carbonate gravity flows and basalt were filled on rifting basins. Meanwhile, the platform was mainly deposited shallow marine carbonate and clastic rocks, and continental tholeiite, reef-building organism developed well. The reefs distributed along the edge of the platform. The upper Yangtze platform was deposited with platform carbonate and evaporite rocks and remained stable in the early and middle Triassic age.

The development of continental foreland basin(T_3–Kz)

As the conversion of rifting basins Dabashan and Chuan-Xiang-Qian fold and thrust belts were formed in the peripheral of Yangtze block in the late Triassic. In the front of foothill the upper Triassic system, which was composed of alluvial fan-braided river conglomerate, sandstone interbedded with shale and coal seam, lay unconformably or disunconformably on the middle Triassic marine sediments. This implies that the continental tectonic basin began to develop in that time.

The west of the basin communicated with open sea and marginal reef and bank, sandbar and delta of marine and continental-marine facies were deposited there in the late Triassic(Fig. 5). Then argillite interbedded with think marl and coal seam of semideep-lake facies and marsh facies were deposited in the latter of Triassic. Longmenshan fold and thrust belt formed with the closure of Ganzi-Litang oceanic basin, and hundreds meter of limeconglomerate accumulated in the front of foothill, at the end of Triassic age. Up to then the four fold and thrust belts developed in the peripheral Sichuan basin. The platform basin formed in the Palaeozoic-middle Triassic age was changed into intracontinental tectonic basin, the Sichuan foreland basin.

From Jurassic to Neogene age Longmenshan fold and thrust belt had been strongly developing. A series of alluvial fans, about 6000--7000m in thickness, were distributed in front of foothill. The depocenter migrated towards southwest of the basin as the development of thrusting from northwest to southeast. There is about 12,000m sediments filled in the foredeep of Longmenshan, with an age of the Sinian to middle Triassic. The platform carbonate rock is about 4000m in thickness, and the thickness of upper Triassic

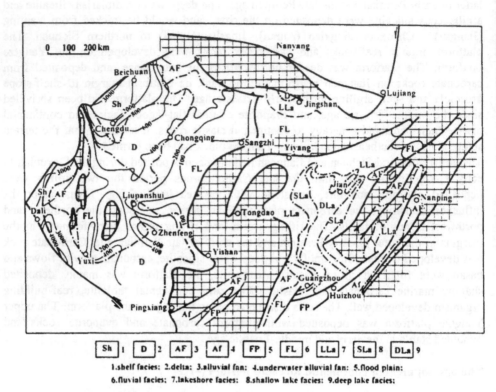

1.shelf facies: 2.delta: 3.alluvial fan: 4.underwater alluvial fan: 5.flood plain:
6.fluvial facies: 7.lakeshore facies: 8.shallow lake facies: 9.deep lake facies:

Figure 5. The lithofacies of the Late Triassic period in the Sichuan foreland basin

to Neogene fluviolacustrine sediments about 8000m. The thickness of sediments reduced towards the central Sichuan foreland basin and formed an unsymmetrical wedge(Fig. 6).

The distribution of lithofacies was evidently controlled by the fold and thrust belts around of the basin. The alluvial fan, alluvial plain, stream, lake shore, delta and lake facies were distributed successively from border to the interior of the basin. The depocentre migrated towards fold and thrust belt of Longmenshan.

The formation of the foreland basin was a result of intracontinental subduction(Fig. 7). It developed in the interior of continent, and not accompanied by ophiolite, melange and volcanic arc. There was a belt of "S" type granite that was parallel to the trend of fold and thrust belt, such as in the hindland of Longmenshan and Qinling fold and thrust belts. This was the result of remelting under the continental crust caused by thrusting and the shortening of deep continental crust so as to balance the shortage of covers in the basin. The basin was only deposited with molasse and formed one-layer filling structure.

Clearly, the type of foreland basin was different from the type-collision foreland basin, and classified as the type of intracontinent subduction foreland basin.

The main structural styles in the Sichuan foreland basin

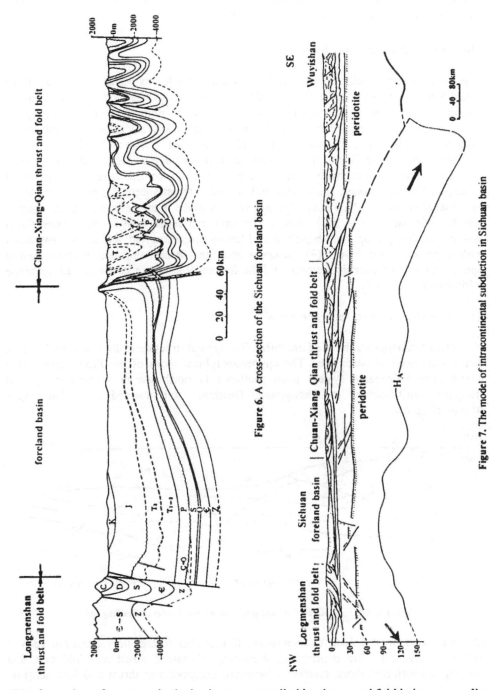

Figure 6. A cross-section of the Sichuan foreland basin

Figure 7. The model of intracontinental subduction in Sichuan basin

The formation of structure in the basin was controlled by thrust and fold belts surrounding the basin. The earliest deformation took place in the Indo-sinian period, but was located in Longmenshan area, such as Kuangshanliang, Tianjingshan, Haitangpu and Zhongba anticlines. Most structures were formed in the early period of the Himalayan orogeny, and all of the sedimentary covers were involved in the deformation.

The structural levels in the basin

The structural deformation shows clearly zonation in the foreland basin. There are three detachment layers in the cover of the basin. They are (1) gypsum and salt layers in the Jialingjiang Formation(fourth member) and the Leikoupo Formation, lower and middle Triassic system; (2) argillite and sandy shale in the lower Silurian system; (3) mudstone, gypsum and salt layers in lower and middle Cambrian system. The deformation is different from in the different structural levels.There are three levels: (1) the lower level: the basement rocks were involved in the deformation, and formed basement detachment zones, ductile shear zones, shear folds and basement thrust sheets, (2) the middle level: the Palaeozoic and middle Triassic systems were involved in the deformation, and formed mainly deformation level in the basin. The structural styles include: imbricated thrust systems, duplex, pop up, fault-bend fold and fault-propagation fold, and show disharmony with upper and lower levels. (3) the upper level: the deformation mainly developed in upper Triassic to Cenozoic systems, and the feature of structure is dominated by gentle folds(concentric folds).

The main structural styles in the basin

(1) Thrust and nappe structure of basement: The typical structure is the Jiangnan-Xuefeng thrust and nappe of basement. The epimetamorphical rocks of the Proterozoic group overthrusted Palaeozoic system from southeast to northwest in western margin of Jiangnan uplift(Leishan to Chongjiang, Guizhou), and formed a folded nappe structure(Fig. 8).

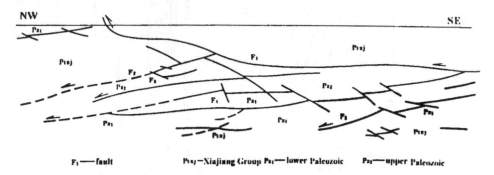

Figure 8. An explanation of seismic profile in western border of Xuefeng uplift

(2) Imbricated thrust system: The imbricated thrusts often formed imbricated fan, such as imbricated fan and blind structure in Wangcang, Dabashan thrust and fold belt, and Haitangpu imbricated thrust system in Jiangyou, Longmenshan thrust and fold belt(Fig. 9).

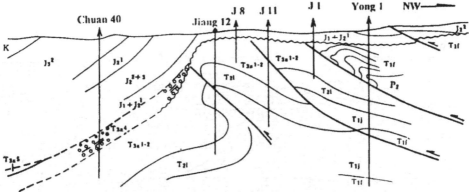

Figure 9. The imbricate thrust system in Haitangpu, Jiangyou,
Longmenshan thrust and fold belt

(3) Nappe structure: The nappe structure is mainly consisted of carbonate rocks of the Palaeozoic system, which often overlaps on the Mesozoic and Cenozoic systems, such as the Pengguan nappe structure groups in the Longmenshan thrust and fold belt(Fig. 10).

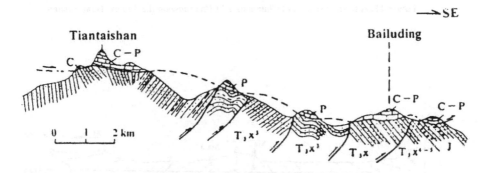

Figure 10. Nappes in Pengxian, Longmenshan thrust and fold belt

(4) Duplex: The duplex is often developed in the Palaeozoic system between the upper glide horizon of the lower Mesozoic system and the lower glide horizon of Palaeozoic system. An example is shown by the Chuanbei-S-11 seismic profile(Fig. 11).

(5) Pop up: The structure is developed well in the basin, such as Shilongxia anticline in the east of Sichuan (Fig. 12), Zhongba anticline and Daxing anticline in the west of Sichuan.

(6) Convergent thrusts and triangular zone: As two thrusts displace in opposite direction, they formed a triangular zone. And the footwall of the thrusts are often deformed a flexure and lead to the disharmonious fold in vertical section ,such as the Tingzipu structure in the west of Sichuan(Fig. 13).

(7) Fault-propagation fold: This structural style is the most developmental one in the basin, such as the Pubaoshan anticline in the east of Sichuan(Fig. 14).

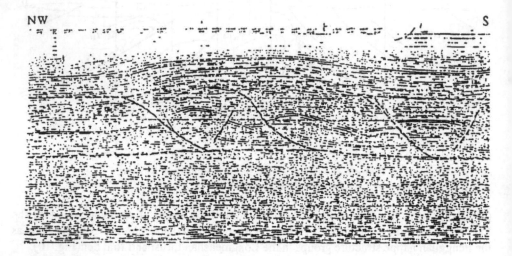

Figure 11. A seismic section of Chuanbei S-11 line showing the duplex, Huayingshan

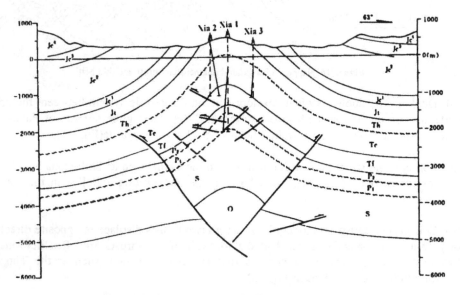

Figure 12. The pop up structure in Shilongxia, Chuandong

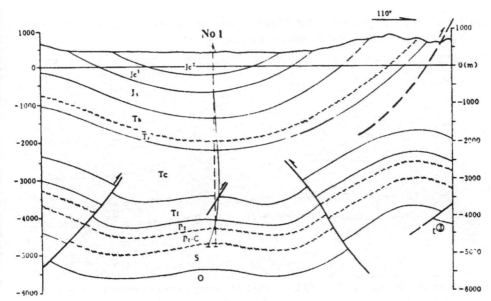

Figure 13. A triangular zone in Tingzipu, Chuandong, showing disharmonic in vertical

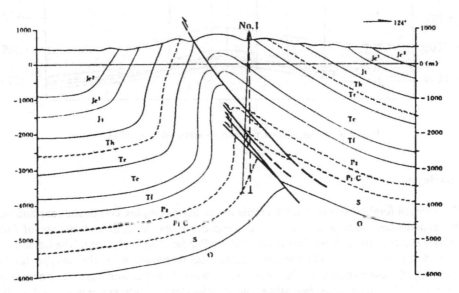

Figure 14. A fault-propagation fold in Pubaoshan, Chuandong

(8) Fault-bend fold: An example is the Wolonghe anticline in the east of Sichuan(Fig. 15).

(9) Gentle fold: The gentle fold has a lower amplitude and fault is not developed, such as the Nanchong gentle anticline in central Sichuan(Fig. 16).

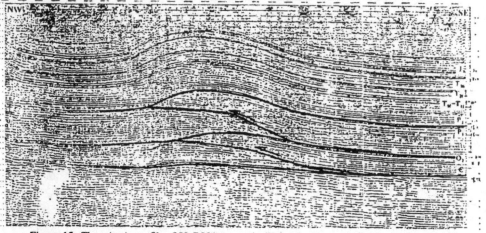

Figure 15. The seismic profile of 83-D331 across the Wolonghe structure, showing fault-bend fold

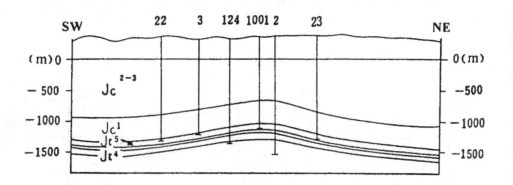

Figure 16. A cross-section of Nanchong gentle anticline in Chuanzhong

Conclusion

The Sichuan foreland basin is not a simple basin, but a complex composite foreland basin that experienced multiple extension and compression by the drift and closure of Proto-Tethys, Palaeo-Tethys, New-Tethys and Palaeo-Pacific. The evolution of the basin can be divided into two stages: (1) the early stage, the formation of collision-type foreland basins(O_2–S), associated with the collision between Yangtze and Huaxia, Yangtze and Huabei blocks; and (2) the late stage, the development of intracontinental subduction-type foreland basin, controlled by the intracontinental subduction taken place since the latter of Triassic age. Therefore, there was different filling structure, sedimentation, magmatic activity, metamorphism in different type foreland basin.

The deformation of covers clearly shows the plane and vertical zonation in the basin. There are different structural styles in different structural levels, and the intensity of deformation gradually weakened towards the basin. The main structural styles include imbricated thrust system, nappe structure, duplex, pop up, triangle zone, fault-bend fold, fault-propagation fold and gentle fold.

References

1. Harding, T. P., and J. D. Lowell, 1979, Structural styles, their plate -tectonic habitats, and hydrocarbon traps in petroleum provinces, AAPG., v.63, p.1016~1056.
2. Beaumont, C., 1981, Foreland basins, Geophys. J. R. Asst. Soc. v.65, p.291~329.
3. Zhong Teqiang, 1982, The prospect and exploration of oil and gas in the Sichuan basin, The Industry of Natural Gas, v.2, p.20~27.
4. Butler, R. W., 1982, The terminology of structure in thrust belts, Journal of structural geology, v. 4, No. 6, p.239~245.
5. Boyer, S. E., and Elliott, D., 1982, Thrust systems, AAPG., v. 66, No. 9, p.1196~1230.
6. Allen, P. A., P. Homewood and G. D. Williams, 1986, Foreland basin: an introduction, in: Allen, P. A., and P. Homewood, ed., Foreland basins: International Association of Sedimentologists Special Publication, No. 8, p.3~12.
7. Mattauer, 1986, Intracontinental subduction, crust-mantle decollement and crust-stacking wedge in the Himalayas and other collision belts, Collision Tectonics, 1986.
8. Heller, P. L., Angevine, C. L., et al., 1988, Two-phase stratigraphic model of foreland-basin sequences, Geology, v.16, p.501~504.
9. Zhang Guowei, 1988, The evolution and formation of Qinling orogenic belt, The Press of Northwestern China University, 1988.
10. Liu Hefu, Liang Huishe and Cai Liguo, et al., 1990, The evolution of the Palaeo-Tethys and the formation and deformation of the superimposed basins in western Sichuan and western Yunnan provinces, in: Wang Hongzhen ed., Tectonopalaeogeography and Palaeogeography of China and Adjacent Regions, China University of Geosciences Press, 1990, p.89~108.
11. Wang Xing and Yang Shufeng, 1992, The tectonic evolution of the early Palaeozoic collision orogenic belt in Southern China, in: Li Jiliang ed., Research of continental and oceanic lithosphere structure in Southeastern China, China scientific technology Press, 1992, p.111~118.
12. Cai Liguo and Liu Hefu, 1993, Characteristics of the Longmenshan fold-thrust belt and foreland basin, in: Sun Zhaocai, Guo Zhengwu ed., Study on oil/gas exploration of Palaeozoic-Mesozoic marine systems in the south China, The Scientific Press, 1993, p.58~71
13. Cai Liguo, Zheng Bing and Liu Jianrong, et al., 1993, The general petroleum geology in the eastern Qinghai-Tibet plateau, The Press of Nanjing University, 1993, p.54~69.

14. Li Jinseng, Cao Xuanduo and Yang Jialu, et al., 1994, The evolution and sedimentation of Qinling oceanic basin in the Phanerozic eon, The Press of Geology, 1994, p.166~185.
15. Liu Hefu, Liang Huishe and Cai Liguo, et al., 1994, The structural styles of Longmenshan thrust-fold belt and the evolution of foreland basin in western Sichuan, Journal of Geology, 1994, v.68, NO.2, p.101~118.
16. Willett S. D., et al. 1994, Subduction of Asian mantle beneath Tibet inferred from models of continental collision, Nature, v.369, No. 23 p.642~645.

Proc 30ᵗʰ Int'l. Geol. Congr., Vol. 6, pp. 105-119
Xiao Xuchang and Liu Hefu (Eds)
© VSP 1997

Contribution to the plate-tectonic interpretation of Sanjiang-Area of Western Yunnan

WANG Yizhao[1], BANNERT, D.[2], HELMCKE, D.[2], INGAVAT-HELMCKE, R.[3], STEINBACH, V.[2], DUAN Jinsun[1], ZHANG Gang[1], BEI Jian[1]

1 - Yunnan Bureau of Geology and Mineral Resources, China.
2 - Federal Institute for Geosciences and Natural Resources, Germany.
3 - Institute of Geology and Dynamics of the Lithosphere, University of Goettingen, Germany.

Abstract

Western Yunnan is characterized by a complicated and long ranging geodynamic evolution. In recent years interest centered on the question where and when terranes of possible Gondwana origin were accreted to the proven continental crust of the Yangtze Platform.

In this contribution published data on the sedimentary evolution of Western Yunnan from Carboniferous to Jurassic times are reviewed critically and new own data on Carboniferous and Permian fossils and sediments, which are strongly climatically controlled, are added.

The geological evolution of Western Yunnan will be compared with the adjacent geological units of Northern Thailand.

Introduction

Yunnan is located to the east of the eastern termination (Eastern Syntaxis) of the Himalaya and hosts the links between the geological units of Tibet with those which characterize the geology of mainland Southeast Asia. Its geodynamic evolution is therefore a key to the understanding of the tectonic history of southwest and central China and neighbouring countries.

During the past decade interest centered on the question which parts of this region might be suspect terranes of Gondwana origin, where the trace of the "Paleotethys" might be located and which regions could be included into the former northern continental margin of Tethys. The question, whether or not this boundary can be traced towards north into the triangle shaped Songpan-Garzi region is of special interest. As possible hosts of this suture, four different zones in Yunnan were discussed in the literature (Fig. 1):

- Ailaoshan Tectonic Belt (including the Red River Fault Zone)

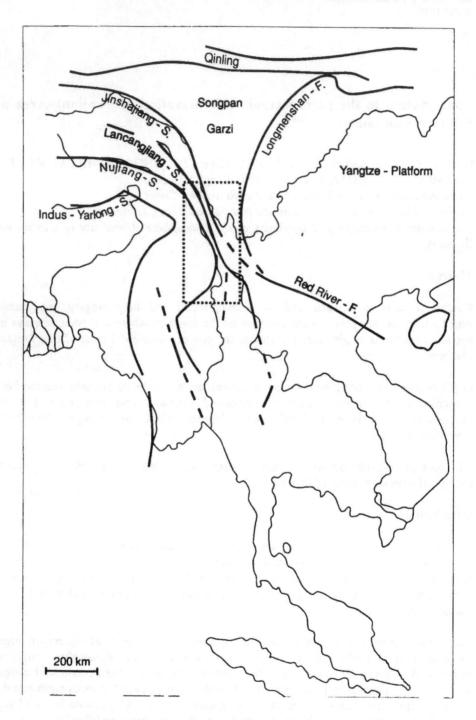

Figure 1: Location of Sanjiang-Area and some important sutures and tectonic lines

- Lancangjiang Fault Zone
- Changning - Menglian Tectonic Belt
- Nujiang Fault Zone.

In this paper we will discuss published data and own new results. They concern the distribution of climatically sensitive faunas and sediments, the distribution of preorogenic, synorogenic and late-orogenic sediments (Fig. 2). Finally, the geodynamic evolution of the Sanjiang-Area will be reconstructed by considering the whole sedimentary, magmatic and metamorphic prozesses, and in comparison to the geology of adjacent areas (Fig. 3).

Yunnan is characterized by strongly varying stratigraphic sections during Paleozoic and Mesozoic times (Fig. 4). The geological evolution of eastern Yunnan is similar to the rest of the Yangtze Platform. It is characterized by Upper Paleozoic strata indicative of warm climatic conditions. On the other hand, the sections from the Baoshan Block and the Tengchong Block were interpreted by several authors for instance by JIN (1994) and WOPFNER & JIN (1996) as indicative of glacial-marine environments during the Upper Paleozoic and were therefore included into the Gondwana realm.

In the following chapters, we will give a brief description of the different blocks and tectonic belts, starting in the west with the Tengchong Block and end in the east with the Ailaoshan Tectonic Belt (comp. Fig. 4 and 5).

Geological evolution of the Tengchong Block

On the Tengchong Block, the formation of possibly glacial-marine origin is known as the Menghong-Group with the Zizhi- and Kongshuhe-Formations.

The Kongshuhe-Formation consists of pebbly mudstone, mudstone and siltstone. The clasts in the pebbly mudstones indicate a source area composed mainly of cristalline rocks (comp. JIN, 1994), which is a significant difference to the pebbly mudstone in the Dingjiazhai-Formation on the Baoshan Block.

From the stratigraphic point of view it is important to note that the overlying limestones of the Dadongchang-Formation contain Triticites sp. (FAN, 1993) of uppermost Carboniferous. But recent information by UENO (written communication, 1996) as well as own preliminary observation indicate that these limestones are somewhat younger. MEISCHNER (IMGP, Göttingen) interpreted these limestones as indicating open platform condition with high energy and shallow, warm water.

This shows that the age of the clastic Menghong-Group might be Stephanian to Asselian, the periods when glaciation on eastern Gondwana peaked. The carbonate production became prominent only during the upper Lower Permian and clastic influx was nil.

Geological evolution on the Baoshan Block

	Gondwana and Gondwana derived terranes	Tethys (Paleotethys)	Northern continents and related terranes
TRIASSIC U			granites, closure of Triassic rifts
M			
L		synorogenic clastics (flysch)	Upper Permian – Triassic rifting
		pelagic sediments	
PERMIAN U	*Shanita*		*Paleofusulina*
	(oolitic limestone)		Emei Shan – basalt
	first limestone	pelagic sediments	*Verbeekina*
M			Middle Permian synorogenic clastics (flysch)
	glacial - marine sediments	pelagic sediments	
L	no fusulinids		connection with artic fauna realm oolitic / oncolitic limestones with fusulinids (*Pseudoschwagerina*)
CARBONIFEROUS U	*Glossopteris* flora		*? Walchia*
M			
L			*Paripteris flora*

Figure 2: Significant features used for the division of the main elements of SE-Asia from Carboniferous to Triassic

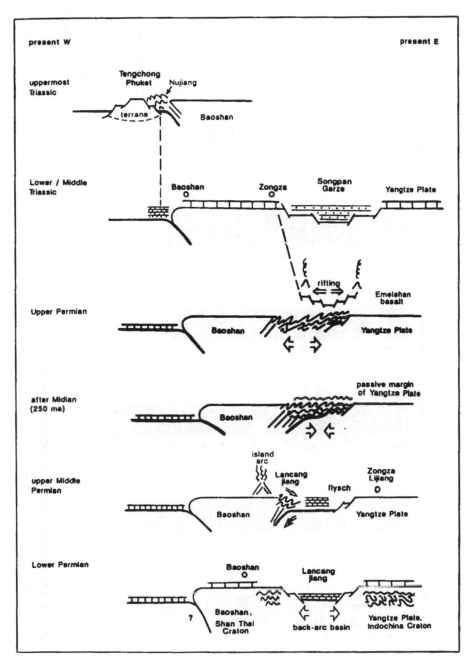

Figure 3: Schematic sketch showing geodynamic evolution of central and north Sanjiang-Area (SW China)

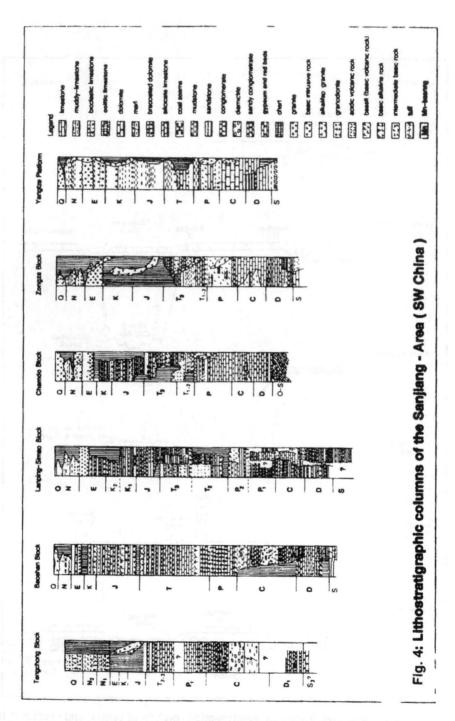

Figure 4: Lithostratigraphic columns of the Sanjiang-Area (SW China)

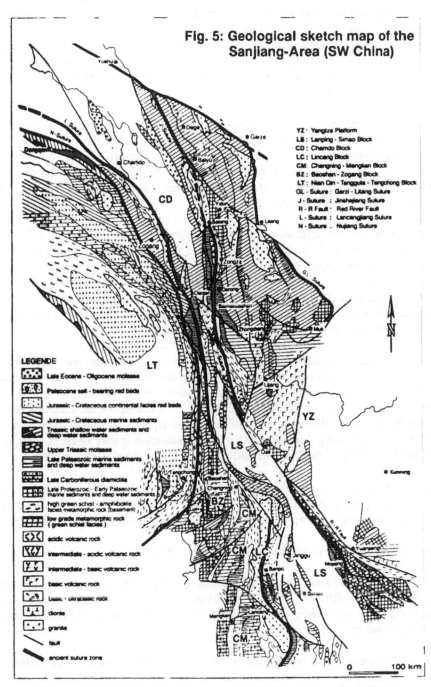

Fig. 5: Geological sketch map of the Sanjiang-Area (SW China)

YZ : Yangtze Platform
LS : Lanping - Simao Block
CD : Chamdo Block
LC : Lincang Block
CM: Changning - Menglian Block
BZ : Baoshan - Zogang Block
LT : Nian Qin - Tanggula - Tengchong Block
GL - Suture : Garzi - Litang Suture
J - Suture : Jinshajiang Suture
R - R Fault : Red River Fault
L - Suture : Lancangjiang Suture
N - Suture . Nujiang Suture

LEGENDE
Late Eocene - Oligocene molasse
Paleocene salt - bearing red beds
Jurassic - Cretaceous continental facies red beds
Jurassic - Cretaceous marine sediments
Triassic shallow water sediments and deep water sediments
Upper Triassic molasse
Late Palaeozoic marine sediments and deep water sediments
Late Carboniferous diamictite
Late Proterozoic - Early Palaeozoic marine sediments and deep water sediments
high green schist - amphibolite facies metamorphic rock (basement)
low grade metamorphic rock (green schist facies)
acidic volcanic rock
intermediate - acidic volcanic rock
intermediate - basic volcanic rock
basic volcanic rock
basic - ultrabasic rock
diorite
granite
fault
ancient suture zone

0   100 km

Figure 5: Geological sketch map of the Sanjiang-Area (SW China)

Seperated from the Tengchong Block by the Nujiang Fault Zone, is the Baoshan Block. On the Baoshan Block the formation of possibly glacial-marine origin is known as the Dingjiazhai-Formation. The age of this formation is said to be Upper Carboniferous to lowermost Permian and overlaps the Tournaisian Pumenqian-Formation. The Pumenqian-Formation is characterized by oolitic limestones (Fig. 6) which indicate a shallow, warm environment. the Dingjiazhai-Formation is built up by sandstone, siltstone, shale, pebbly mudstone (Fig. 9) and contains lenses of bioclastic limestone. The majority of the pebbles in the Dingjiazhai-Formation consists of limestones which are from the Pumenqian-Formation (comp. Fig. 6 and 7).

The stratigraphic range of the Dingjiazhai-Formation is limited by finds of Triticites sp. in the uppermost parts as well as in sedimentary layers intercalated into the roofing basalts of the Woniusi-Formation. Triticites indicates uppermost Carboniferous. This information confirms that the Dingjiazhai-Formation is entirely of Upper Carboniferous age and therefore older than the maximum glaciation in eastern Gondwana which is usually said to be of Stephanian to Asselian age.

This coincides with the following arguments: On top of the Woniusi-Basalt follows the Bingma-Formation which is characterized by red fluviatile sandstones and the Yongde-Formation. On the southern Baoshan Block, a limestone with indications of beginning of formation of oolites was collected which contains Pseudoschwagerina sp. (Fig. 8). Pseudoschwagerina indicates Asselian age. This confirms that at least the southern parts of the Baoshan Block were situated during Asselian times in a warm and arid climate and were inhabited by a fauna with typical Tethyan affinities.

There is no important tectonic fault between this outcrop of the warm water limestone and the possible glacial-marine deposits of the Dingjiazhai-Formation. The given informations (occurence of red continental sandstones and warm water limestones with Pseudoschwagerina) can only be explained by an extreme rapid dislocation of the Baoshan Block from a glacial-marine environment in Upper Carboniferous to an arid and warm environment during Asselian. Such a rapid dislocation is very unusual, therefore the glacial-marine setting of the Dingjiazhai-Formation has to be critically reassessed.

Paleogeographic conclusions should not only be drawn from the distribution of the diamictites (= pebbly mudstones) but it should also taken into cosideration the whole stratigraphic and paleogeographic development from Upper Paleozoic to Triassic times.

Geological evolution in the Changning-Menglian Belt

To the east of Kejie Fault the Carboniferous and Permian strata of the Changning-Menglian Belt show different features. The underlying Devonian strata are homogeneous and indicate deep water sediments. In Carboniferous and Permian different sedimentary environments occur:

- The western sedimentary facies is build up by basic volcanic rocks followed by fossiliferous carbonates with close affinities to the Tethyan fauna. These limestones indicate a shallow and warm environment of an open platform.

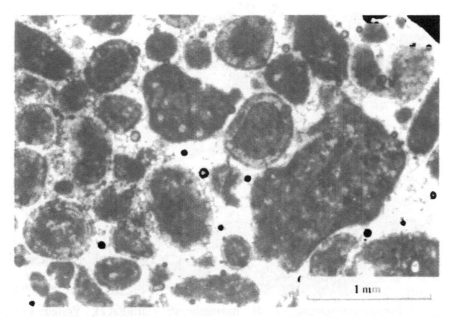

Figure 6: Oolithic grapestone, Pumenqian- Formation (Upper Viseán), Baoshan Block, south of Baoshan, 10 km south of Shuichang

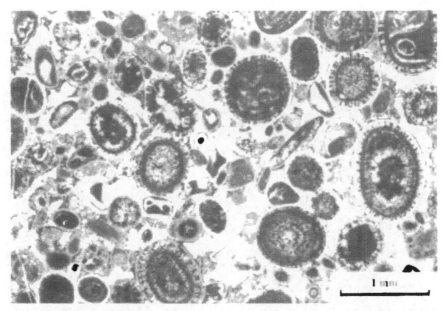

Figure 7: Oolithic grapestone as a clast in the pebbly mudstone, Dingjiazhai-Formation (Upper Carboniferous), Baoshan Block, southwest of Baoshan, Dongshanpo, 4 km east of Youwang

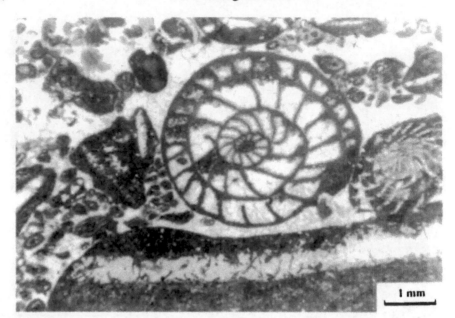

Figure 8: Pseudoschwagerina sp. cf. moungthensis (DEPRAT), Yongde-Formation
(lowermost Permian), Baoshan Block, 8 km west of Zhengkang

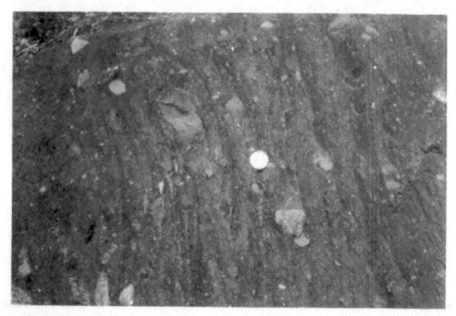

Figure 9: Outcrop of pebbly mudstone, Dingjiazhai-Formation (Upper Carboniferous),
Baoshan Block, 10 km east of Liuku

- The eastern facies is characterized by sandstones and shales.

Because the underlying sequence of this belt is not clear, at least 3 different interpretations of its tectonic setting were proposed:

1) FAN & ZHANG (1993) interpretate this zone as a back arc basin, where the volcanic rocks formed relative with the rifting.

2) LIU (1989), HE & LIU (1993) and others expressed the view that this belt represents the remains of Paleotethys, with characteristics of polyisland ocean.

3) But most geologists, as HUANG et al. (1987) interprete this belt as a suture line.

The warm water environmental character of the sediments and the Tethyan fauna of this belt are not disputed.

Similar sections are also known from Northern Thailand - i. e. Chiang Dao - which are all situated within a region with warm water sediments of Carboniferous and Permian age (comp. FONTAINE et al.,1993). CARIDROIT (1993) argues that such a setting is a typical deep marine environment - which is later involved in nappe and thrust sheets. For the Changning - Menglian Belt HE & LIU (1993) also proposed nappe thrusting from W towards E.

Being aware of the complexity of the Changning-Menglian Belt we hesitate to comment much. Much will depend on the interpretation of the chemistry of the volcanic sequences and there are doubts arising about their stratigraphic age. Since MACDONALD & BARR (1993) proposed that "metamorphic core complexes" might be developed in Northern Thailand (i. e. Doi Inthanon) also the effects of late orogenic extension in the Changning-Menglian Belt (i. e. Mengxin) should be considered.

Lancangjiang Fault Zone

The Lancangjiang Fault Zone is an impressive suture zone. Along this suture zone a Permian flysch-sequence is known with good exposures near Lancang and Deqen (Fig. 11). North of Deqen the folded Permian flysch is covered by Upper Triassic continental red beds with an angular unconformity. The sediments - deep marine siliciclastics, which have been deposited by turbidity currents - witness the well established Lancangjiang orogeny. They can be correlated with similar sequences known from central Thailand, Phetchabun, Nan-Uttaradit (HELMCKE & KRAIKHONG, 1982) and also from regions north of Yunnan, i. e. Karamiran Pass in the Kunlun Shan.

In the same area, south of Deqen, is an outcrop with Permian radiolarian chert (Fig. 10) and serpentinite. According to these evidences - Permian radiolarian chert and serpentinite, folded Permian flysch, Upper Triassic continental red beds and the ungular unconformity - and the geological evolution of Central Thailand (HELMCKE & KRAIKHONG, 1982; HELMCKE, 1985), the Lancangjiang Suture closed during

Figure 10: Ribbon chert (Lower Permian), 25 km north of Benzela, northwest Yunnan

Figure 11: Flysch-sequence, overturned; (Permian); on the Lancangjiang, 42 km north of
          Deqen, northwest Yunnan

Permian and has to be interpreted as the spur of a marginal ocean, probably a back arc basin. This marginal ocean divided the Yangtze Platform proper from the western adjoining regions untill they (re-) united during Permian times.

For two different reasons the Lancangjiang Suture is unlikely to represent the spur of a "Paleotethys"-ocean:

> - the warm water nature of the Lower Permian sediments to the east and the west of the Lancangjiang Suture calls for a marginal ocean, and

> - the closure of the Lancangjiang Suture took place during the Middle Permian, much too early to accomodate the drift from Gondwana to the southern Eurasia shores.

After closure of this suture during Middle to Upper Permian times rifting occured again along the Lancangjiang. This is documented by thick Triassic volcanic sequences and new basins came into existence. One example of this process is exposed in the Bai Ma Xue Shan mountains: here Triassic basalts (with pillow structures) are associated with clastics and shallow marine limestones. This situation equals much the situation in the Lampang area of Northern Thailand.

Ailaoshan Tectonic Belt

Stratigraphy, facies evolution and fossil content of the strata to the east of the Lancangjiang Fault Zone are so similar to the evolution of the Yangtze Platform proper during Upper Paleozoic times that they hardly provide any argument to assume that the Ailaoshan Tectonic Belt can be regarded as an important divide between both areas. During the Himalayan stage strong deformation overprinted the Lanping - Simao area, especially the Ailaoshan. But even the effects of the overprint, which is obviously seen along the Red River Fault, are disputed: while TAPPONIER et al. (1986) favour strong eastward lateral extension of Tibet and SE-Asia, DEWEY et al. (1989) conclude that only little lateral extrusion or escape occured.

Nujiang Fault Zone

On the Triassic strata of Western Yunnan it must be added that there is one narrow strip of Middle Triassic siliciclastic sediments between Luxi and Ruili. These siliciclastics are laid down by turbidity current in that zone which has to be regarded as part of the Nujiang Fault Zone. Also one small outcrop of ribbon cherts (age unknown) and serpentinites is known here. The mentioned siliciclastics are labelled flysch in some papers and own field-observations do not disagree.

The Nujiang Fault Zone seperates the Baoshan Block in the east and the Tengchong Block in the west. According to the proven warm water fauna (Pseudoschwagerina) and the warm climatical sediments (i. e. red bed sediments in the Yongde-Formation) on the

Baoshan Block and the Middle Triassic siliciclastic turbidites, we assume, that the Nujiang Fault Zone might be an important suture between a Gondwana terrane in the west and the margin of Eurasia in the east. Further investigations on this subject are neccessary.

The complex geology of western Yunnan not only reflects the complicated evolutionary history during Paleozoic and Mesozoic - with an old Variscan orogeny in the east, an Middle Permian orogeny along the Lancangjiang Fault Zone and an Middle to Upper Triassic orogeny along the Nujiang Fault Zone - but also includes the deformation and displacement during the Himalayan stage.

Literaturverzeichnis:

1.   Caridroit, M., 1993. Permian radiolarian from NW Thailand. *Proc. Intern. Symposium on Biostratigraphy of Mainland Southeast Asia: Facies and Paleontology (BIOSEA)*, v.1, 83 - 96.

2.   Dewey, J.F., Cande, S., Pitman, W.C., III, 1989. Tectonic evolution of the India/Eurasia collision zone. *Eclogae Geologicae Helvetiae*, 82, 717 - 734.

3.   Fan Chengjun, Zhang Yifei, 1993. On the Structural Pattern of Western Yunnan. *Geological and Mineral Resources Bureau of Yunnan.* 12 p.

4.   Fontaine, H., Suteethorn, V. and Vachard, D., 1993. Carboniferous and Permian limestone in Sop Pong area: unexpected lithology and fossils. *Proc. Intern. Symposium on Biostratigraphy of Mainland Southeast Asia: Facies and Paleontology (BIOSEA)*, v. 2, 319 - 336.

5.   He Fuxiang, Liu Benpei, 1993. Recognition of Ancient Oceanic Island in Paleo-Tethys, Western Yunnan. *J. of China Univercity of Geosciences*, 4/1, 23 - 29.

6.   Helmcke, D. and Kraikhong, C., 1982. On the geosynclinal and orogenic evolution of the Central and Northeastern Thailand. *J. Geol. Soc. Thailand*, 5, 52 - 74.

7.   Helmcke, D., 1985. The Permo-Triassic "Paleotethys" in Mainland Southeast-Asia and adjacent parts of China. *Geol. Rdsch.*, 74/2, 215 - 228.

8.   Huang Jiqing, Cheng Bingwei, 1987. The Evolution of the Tethys in China and Adjacent Regions. *Geol. Publ. House*, Beijing, 109 p.

9.   Jin, X., 1994: Sedimentery and Paleogeographic Significance of Permo-Carboniferous Sequences in Western Yunnan, China. Ph.D., *Geol. Inst. University of Cologne, Sonderveröffentlichungen*, 99, Köln, 136 p.

10. Liu Benpei, Feng Qinglai and Xie Xuewen, 1989. Tectono-Paleogeography and Paleo-Biogeography of the Changning-Menglian Belt in Western Yunnan during Hercynian-Indosinian Stage. *Reports and Abstracts 4th Intern. Symp. Pre-Jurassic Evol. East Asia. 1,* 14 - 17.

11. Macdonald, A.S., Barr, S.M., Dunning, G.R. & Yaowanoiyothin, W., 1993. The Doi Inthanon metamorphic core complex in NW Thailand: Age and tectonic significance. *Journ. S.E. Asian Earth Sci.,* v. 8, 117 - 126.

12. Tapponnier, P., Peltzer, G. and Armijo, R., 1986. On the mechanism of the collision between India and Asia. *Collision Tectonics,* M.P. Coward and A.C. Ries (editors), *Geol. Soc. London, Spec. Publ.,* v. 19, 115 - 157.

13. Wang Yizhao, Xiong Jiayong, Lin Yaoming, 1988. Some features on Yunnan geological structures. *Regional Geology of Yunnan.* v. 7, 105 - 111. (in Chinese with English abstract).

14. Wopfner, H. and Jin, X., 1996. Late Palaeozoic glaciomarine deposits of West Yunnan and their Gondwana origin. *Geowissenschaften,* 14, 288 - 290.

Proc 30ª Int'l. Geol. Congr., Vol. 6, pp. 121-132
Xiao Xuchang and Liu Hefu (Eds)
© VSP 1997

Lateral Variation of Crustal Structure from Himalaya to Qilian and Its Implication on Continental Collision Process

Zeng Rongsheng   Ding Zhifeng   Wu Qingju
Institute of Geophysics, SSB, Beijing 100081, China

Abstract

In the Sino-US joint project of Tibetan Plateau in 1991-1992, 11 broadband digital PASSCAL instruments were deployed. Of them, eight instruments were arranged along the Lhasa-Golmud profile, nearly in N-S direction. The results of receiver function from the teleseismic events clearly indicate the dipping Moho in the Tibetan Plateau and an offset of Moho about 15km occurs between Amdo and Wenquan, adjacent and north to the Bangong-Nujiang ancient suture. It is shown that the variation of the arrival time of Pn from local events confirms the results from the receiver function. An independent result from the seismic refraction profile in western Tibet displays also an offset of Moho of about 10 km adjacent and north to the Bangong-Nuijiang suture. Therefore, it is reasonably suggested that the crustal thickness in Qiangtang block has been reduced by 10-15 km after its thickening during the collision of Lhasa block with Qiangtang block due to the gravitational isostatic effect.

The crustal structure from Himalaya to Qilian indicates reverse dipping of Moho. Reverse dipping may also exist in the intra-crustal structure. It shows that the compression and also the penetration from both southern and northern boundaries have to be considered in the continent-continent collision process.

Keywords: Himalaya, Qilian Mountain, Crustal Structure, Collision Process

1. The lateral variation of crustal velocity structure from Himalaya to Qilian Mountain

(a) The results from receiver function
The lateral variation of crustal structure from Himalaya to Qilian Mountain is the key topic for the investigation of India-Eurasia collision process. In the last few years, several joint groups have made important progress in the investigation of crustal structure in Tibetan Plateau. The vertical reflection of Moho is very clear underneath Himalaya but it is not observed near   and north to the Yaruzampbo suture. However, the regional phase Pn from local seismic events and the P-S converted waves of Moho from teleseismic events were observed nearly everywhere in the plateau.

From July 1991 to June 1992, 11 broad-band digital PASSCAL instruments were deployed inside the Tibetan Plateau (Fig.1) according to a Sino-U.S. joint project [1,2]. The stations are abbreviated as:

Xiga: Xigatze   Lhsa: Lhasa
Sang: Sangxiong   Amdo:Amduo
Wndo: Wenquan   Erdo: Erdaogou
Budo: Budongquan   Tunl: Golmud
Gang: Linzhi   Ushu: Yushu
Maqi: Magin

Most of them are located on the main road from Lhasa to Golmud. About 100 local events and total amounts of 700 seismic events were recorded. They provide an excellent possibility to investigate the lateral variation of velocity structures between different parts inside the Plateau.

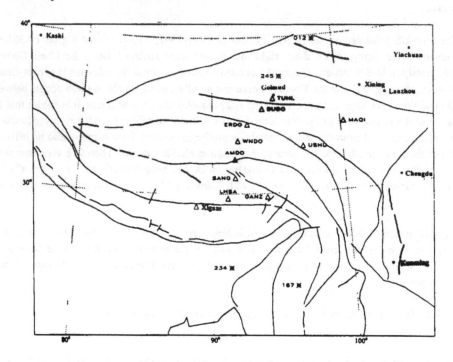

Figure 1. Location of digital seismic stations and local events. Triangles indicate digital seismic stations; open circles, local events; dots, the place names; lines, geological structures.

The crustal velocity structures underneath the eleven stations were obtained by the inversion of teleseismic waveforms in broadband seismic recordings. Within the first 20-30 seconds of the teleseismic seismogram, usually it contains different P-S converted waves from different layers in the lithosphere underneath the station. The converted wave from Moho is the strongest. It provides a convenient method especially for the investigation of the lateral variation of lithospheric structures in some inaccessible places like Tibet. Another advantage of this method is that the incident ray from teleseismic event is nearly vertical, it minimizes the effects due to the lateral variation of the velocity structure in the neighboring area. The receiver function

$$R(t) = \int \frac{R(\omega)}{Z(\omega)} e^{i\omega t} d\omega$$

is introduced where R(ω) denotes the radial horizontal component, and Z(ω) the vertical component of the seismogram. In the process of deconvolution, it removes the source and propagation effects, and the lithospheric structures underneath the station can be obtained [3-5].The noise due to random disturbances can be effectively reduced by stacking the receiver functions calculated from events clustered in a small range of epicentral distances. In some favorable conditions, inhomogeneity of the structure beneath the station can be investigated from the difference of receiver functions obtained from events in different azimuth [6].

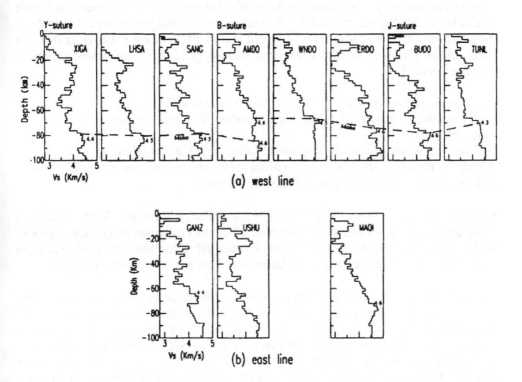

Figure 2. Crustal velocities and Moho depth in Tibetan Plateau from receiver function. The Moho underneath each station was marked as the S-wave velocity reaches or exceeds 4.4 km/sec. Lateral variation of the Moho depth is shown by dash line.Y indicates Yaruzampbo suture; B, Bangong suture; J, Jingsha suture.

Receiver functions in Tibetan Plateau were calculated either in frequency domain [7] or time domain [8]. Wu used different methods of filtering in order to reduce the random noises more effectively. Inversion of the receiver functions from Wu for a one-dimensional S-wave velocities beneath the 11 stations in Tibet is shown in Fig.2 [8-10].

The velocity structures in Xiga, Lhsa, Sang, Amdo,Wndo, Erdo, Budo and Tunl are arranged in a profile (west line)from south to north. Another profile (east line) includes Gang, Ushu and Maqi.

A zone with very low velocity is located at a depth from 10 to 20 km underneath Xiga, Lhsa, Sang, Amdo. The low velocity zone becomes deeper underneath Erdo and Budo. An intra crustal discontinuity at a depth from 40-50 km is observed also.
Moho discontinuities are determined as the S-wave velocity reaches or exceeds 4.5 km/sec, except in Xiga and Tunl where velocity of 4.4 km/sec or 4.3 km/sec are observed respectively. The S-wave velocity of 4.5-4.6km/sec in the uppermost mantle is similar to the Sn velocity we have obtained from local travel times. Usually, a transition layer of 5-8 kilometers is observed at the bottom of the lower crust from the inversion of receiver functions, as Hirn had reported from his investigation of refraction data [11].

Because the converted waves P-S from Moho is very strong, the lateral variation of Moho depth is considered to be more reliable. The main features of Moho are summarized as:
(i) The north dipping Moho is observed in Lhasa and Qiangtang blocks;
(ii) Abrupt change at least of 10 km in Moho depth occurs between Amdo and Wndo;
(iii) The Moho dips southerly from Golmud to Erdo.

We can see also that the crustal S-wave velocity beneath Amdo increases monotonously from 3.5 km/sec at a depth of 23 km to 5.0 km/sec at a depth of 85 km. Moho depth becomes shallower in North Central Tibet had been reported also from DSS investigation [12], surface wave dispersions [13] and converted waves [14].

(b) Confirmation of the Moho depths from receiver function by the travel times of Pn
In order to confirm the lateral variation of Moho depth along Lhasa-Golmud profile obtained from the receiver function, we investigate the first arrivals of the stations between Lhasa and Golmud from the two local events 92012 and 91245 at the north to Golmud, and another two local events 91234 and 92167 at the south to Lhasa (Fig.1). Their locations are nearly on the same line of the seismic stations on the main road from Lhasa to Golmud. They constitute as an ideal observational system to discriminate the lateral variation from vertical variation of velocity structure and to detect the inclination of the discontinuity.

Usually, the travel times of Pn for a local event in the Tibetan Plateau are not on a straight line, the reported velocities of Pn in the Tibetan Plateau differ a lot, they vary from 7.8 km/sec to 8.5 km/sec. The irregular distribution of the travel times of Pn may be produced by:

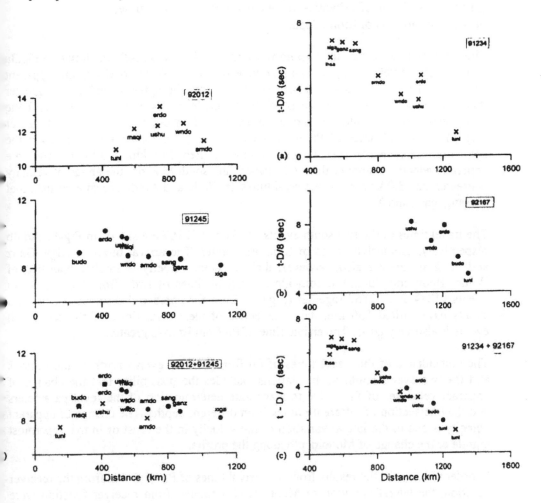

Figure 3. The travel times of Pn from the two north local seismic events 92012 and 91245. a), the travel times from event 92012; dot indicates the travel time for each station. b), the travel times from event 91245; cross indicates the travel time for each station. In (c), the ordinate for (a) is shifted 340 km, so that the travel times of Erdo from the two events coincide.

Figure 4. The Travel times of Pn from the two south local seismic events 91234 and 92167.(a), the travel times from event 91234; cross indicates the travel time for each station. (b), the travel times from event 92167; dot indicates the travel time for each station. In (c), the ordinate for (b) is shifted 150 km, so that the travel times of Erdo from the two events coincide.

(1) Inaccuracy of the determination of the focal parameters;
(2) Intersection of different seismic phases from different depths;
(3) Lateral variation of velocities in the crust or uppermost mantle;
(4) Lateral variation of Moho depth.

The travel time curves of the two north events 92012 and 91245 are shown in Fig.3a and 3b respectively. They are similar to each other. North to Erdo, the apparant velocities of Pn is less than 8.0, and south to Erdo, the apparant velocity is greater than 8.0 km/sec. The arrival times of Wndo and Amdo are earlier. If we shift the coodinates, and coincide the arrival times of Erdo from the two events together (Fig3c), the travel times of Pn from the two events coincide pretty well, even the epicentral distances of the two events have shifted 340 km. North to Erdo, the apparant velocity is smaller than 7.8 km/sec, and south to Erdo, the apparant velocity is greater than 8.0 km/sec. The arrival times of Wndo and Amdo are smaller than that of Sang, ganz and Xiga.

The travel times of the two south events 91234 and 92167 are shown in Fig.4a and 4b respectively. The arrival times of Erdo are greater. The arrival times of Xiga, Ganz and Sang are greater also, however, their apparant velocity is smaller than that of Wndo, Budo and Tunl. We coincide the arrival times of Erdo from the two south events 91234 and 92167 together (Fig.4c). Even the epicentral distances of the two events have shifted 150 km, the coincidence of the arrival times of the two south events is also very good. The arrival time of Pn from Erdo is greater.

The coincidence of the travel times of Pn from the two events north to the stations and the two events south to the stations excludes the possibility that the change of apparant velocities of Pn is due to inaccurate determination of the focal parameters and the intersection of different phases from different depths. The change of apparant velocity is due to the lateral variation of the velocity in the crust or in the uppermost mantle or the change of Moho depth along the profile.

In order to compare the results from the arrival times of Pn with that from the receiver function, the lateral variation of Moho depth obtained from receiver function were used as a preliminary model, and the travel times of the stations between Lhasa to Golmud from the local events were calculated. The mean crustal velocity is assumed to be 6.3 km/sec, and the velocity in the uppermost mantle is adopted to be 8.2 km/sec. Only the differences of the travel times between stations are concerned in this case. For the two north event, the coincidence of the results from receiver function and Pn is very well (Fig.5a,b), except the arrival time of Tunl from event 92012. The greater arrival time of Erdo is produced by the southerly dip of Moho from Golmud to Budo. For the two south events(Fig.5c,d), the coincidence is also very well, the greater arrival time of Erdo is produced by the northerly dip of Moho from Wndo to Erdo. The anomalous arrival times of Pn for some stations can be interpreted properly by this way.

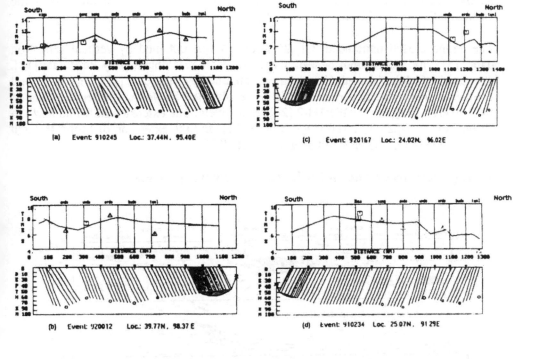

Figure 5. Comparison of the results from receiver function and Pn. In the lower figure, small open circles indicate the Moho depths obtained from receiver function. Stars indicate the location of seismic events, thin lines indicate the seismic rays. In the upper figure, solid lines indicate the calculated travel time curve of Pn by seismic ray tracing. The triangles indicate the observed travel times of Pn from local seismic events.

(c) The crustal structure at the boundaries of Tibetan Plateau

Altyn-Tagh and North Qilian are considered as the northern boundary of Tibetan plateau. Recently, a 100km vertical reflection profile was conducted in the north Qilian in 1993 by the Ministry of Geology of China [15] (Fig.6). A south-dipping structure from the upper crust to Moho is observed. It may be extended and meets the thrust fault on the surface. The south-dipping structure intersects the Moho where it has just been deepened from 41 km to 48 km. A south-dipping thrust was also observed by Tapponnier et al. in Qilian west to this reflection profile [16].

The Moho beneath Himalaya undergoes a break according to the results from Sino-French refraction profile.

The crustal structure from Himalaya to Qilian is shown schematically in Fig.7. From Yaruzampbo suture (YS) to Kunlun, the variation of Moho depth was obtained by receiver function and confirmed by the arrival times of Pn. The Moho depths from receiver function are reduced proportionally in order to be compatible with the vertical reflection profile of INDEPTH-1 [17].

The Moho crustal underneath Qilian mountain was obtained by vertical reflection data [15]. The Moho depth beneath Chaidam Basin was obtained by PmP data [18]. The break of Moho beneath Himalaya was obtained by PmP data [19]. The MHT and another dipping structure north to YS were obtained by vertical reflection data of INDEPTH-1 [17] and INDEPTH-2 [20] respectively. The extension of Moho north to Qilian was obtained by a refraction profile [21], and the extension of Moho through Himalaya to India was matched by gravity data [22].

2. Implications on the India-Eurasia collision process

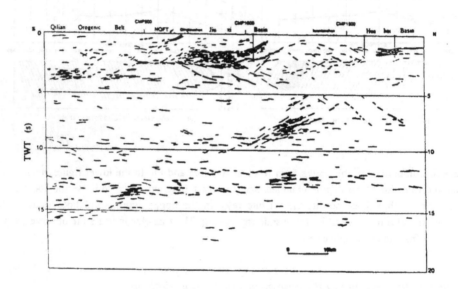

Figure 6. Vertical reflection profile in North Qilian. The prominent south-dipping structure is observed in the middle crust. It can be extended to the Moho where its depth changes from 41 km at the north to 48 km at the south.

(a) The mechanism of crustal thickening in the mountains
The lateral variation of crustal structure from Himalaya to Qilian Mountain (Fig.7) indicates some similarities between Himalaya and Qilian. Both the crustal thrust and crustal thickening are observed in Qilian and Himalaya. The reflection profile of Qilian (Fig.6) shows that the crustal thickening occurs just at the intersection of the intra-crustal thrust and Moho. The regional compression gives rise to both the crustal thickening and the intra-crustal thrust, which in turn produces the uplift on the surface. Most people are convinced that the crustal shortening is one of the principal effects in

continent-continent collision process. Fig.6 provides an excellent example how the intra-crustal thrust, surface uplift, crustal shortening and Moho deepening are linked together and produced by the application of regional compression field.

The crustal structure in Himalaya is not so clear as in Qilian due to insufficient geophysical data. A break of Moho of 18 km underneath the Himalaya in Fig.7 can be interpreted by the effect of crustal thickening, while it is only 7 km underneath Qilian. From the investigation of the seismicity and focal mechanism in Himalaya, Barazangi et al. [23] interpreted the MCT, MBT and MFT as successive thrust faulting. The processes in Himalaya and Qilian are similar in the principal respects.

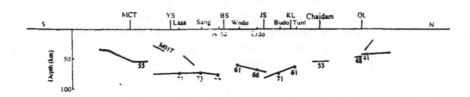

Figure 7. Lateral variation of crustal structure from Himalaya to Qilian. The number indicates the Moho depth. From Lhsa(Lhasa) to Tunl(Golmud), the values of depth were obtained from receiver function. The original values o Figure 2 were reduced proportionally in order to be compatible with the result of Moho reflection from INDEPTH-1. This result was confirmed by the travel times of Pn from local seismic event. The value underneath QL(Qilian mountain) was obtained by the vertical reflection profile. The Moho depth underneath Chaidam Basin was obtained by wide-angle reflection. The Moho depth underneath MCT was obtained by the wide-angle reflection. Dipping structures in the crust indicate the prominent crustal thrusts in different region, obtained from vertical reflections.

(b) The process of penetration
The MHT (Main Himalyan Thrust) in Fig.7 may be interpreted as the upper boundary of the penetrating Indian crust [17]. However, the MHT was not observed further north [20]. Clear reflection of Moho was observed also beneath MHT, but not further north. The Moho depth from PmP changes a little across the Yaruzampbo suture [11]. Variation of the crustal and mantle properties may exist across the YS according to the recent INDEPTH-2 experiment [24]. Some people suggested that the penetrating Indian crust may stop at YS [25]. In this case, a subducting slab of Indian upper mantle may extend into the Euraslan asthenosphere. Another intra-crustal north-dipping structure, similar to that of MHT, was observed beneath Gangdese Batholith. It may be interpreted as another upper boundary of the penetrating Indian crust at the beginning of continent-continent collision, since successive penetrations may have been occurring at different belts on the surface in different geological epoch.

Some evidences indicate that the Indian crust and/or the northern Asian block has been penetrated into the Tibetan lower crust as suggested by Zhao and Morgan [26]. Determination of Q-value in Tibetan Plateau from seismic surface waves [27,28] and the cluster of earthquakes at different depth ranges of 15 km, 33 km and 100 km highly suggests that the Tibetan lower crust has been reworked and the process is still operating at the present time.

(c) The reverse dipping Moho at Jingsha

Recently, some people suggested that the subduction in the northern boundary of Tibetan Platean may also play an important role in the collision process [29,30]. From Fig.7, we observe the north-dipping Moho in Qiangtang block and the south-dipping Moho in Kunlun block. They meet nearly at the ancient Jingsha suture. The reverse dipping Moho indicates that both the compression from India and Mongolia are playing important role in the collision process. It is interesting to notice that the recently published refraction profiles from Alps [31] and Dabieshan of eastern China [32] also show the similar reverse dipping Moho. It is probably that the reverse dipping Moho is a typical structure of the Mesozoic suture.

(d) Explanation of the shallower crustal thickness in Qiangtang

An offset of Moho about 15 km was indicated by the receiver function. It is confirmed by the travel times of Pn. An independent result from the seismic refraction profile in western Tibet displays also an offset of Moho of about 10 km adjacent and north to the Bangong-Nuijiang suture. Therefore it is reasonably suggested that the crustal thickness in Qiangtang block has been reduced by 10-15 km relative to the Lhasa block. We assume Qiangtang block had been thickened about 75 km during the collision of Lhasa block with Eurasia in 1300 M.A.. The reduction of crustal thickness as we observe now is due to the gravitational isostatic effect after its previous thickening.

References

1. Rongsheng Zeng et al. An introduction to the Sino-US joint project "Lithospheric structure and dynamics in Tibetan Plateau", Acta Seism. Sinica 6(2), 249-250 (1993).
2. T.J. Owens et al. PASSCAL instrument performance during the Tibetan Plateau passive seismic experiment, Bull Seism. Soc. Am. 83(6), 1959-1970 (1993).
3. T. J. Owens, G. Zandt, S. R. Taylor. Seismic evidence for ancient rift beneath the Cumberland Plateau, Tennessee: a detailed analysis of broadband teleseismic P waveforms, J. Geophys. Res. 89, 7783-7795 (1984).
4. G.E. Randall. Efficient calculation of differential seismograms for lithospheric receiver functions, Geophys. J. Int. 99, 469-481 (1989).
5. C.J. Ammon, G. E. Randall, G. Zandt. On the nonuniqueness of receiver function inversions, J. Geophys. Res. 95(B10), 15303-15318 (1990).
6. Lupei Zhu, T.J. Owens, G.E. Randall. Lateral variation in crustal structure of the Northern Tibetan Plateau inferred from teleseismic receiver functions, Bull. Seism. Soc. Am. 85,1531-1540 (1995).

7. Lupei Zhu et al. Preliminary study of crust-upper mantle structure of the Tibetan Plateau by using broadband teleseismic body waveforms, Acta Seism. Sinica 6(2), 305-316 (1993).

8. Qingju Wu. Waveform inversion from teleseismic events and lithospheric structures in Tibetan Peateau, Ph.D. thesis, Institute of Geophysics, State Seismological Bureau of China, Beijing (1996).

9. R. Z. Zeng and Q. Wu, A review on the lithospheric structure in the Tibetan Plateau and constraints for dynamics, PAGEOPH 145(3/4), 425-443 (1995).

10. Qingju Wu et al. Receiver function and Tibetan lithospheric structure, in "Structure of the Lithosphere and Deep Process", Proc. 30th IGC (1996).

11. A. Hirn, G. Jobert, G. Witlinger, Z.X. Xu, & E.Y. Gao. Main features of the upper lithosphere in the unit between the high Himalayas and the Yarlung Zangbo Jiang suture, Annales Geophysicae 2, 113-118 (1984).

12. D.Y. Lu and X.J. Wang. The crustal structure and deep internal processes in the Tuotuohe-Golmud area of the north Qinghai- Xizang Plateau, Bull. Chinese Academy of Geol. Sci. 21, 227-337 (1990).

13. C. Brandon, B. Romanowicz. A"no-lid"zone in the central Chang-tang platform of Tibet; Evidence from pure path phase velocity measurements of long period Rayleigh waves, J. Geophys. Res. 91, 6547-6564 (1986).

14. G. Herque, G. Wittlinger and J. Guilbert. Anisotropy and crustal thickness of Northern-Tibet. New constraints for tectonic modelling, Geophys. Res. Lett. 22(14), 1925-1928 (1995).

15. Xuan-Zhi Wu et al. Research on the fine crustal structure of the northern Qilian-Nexi Corridor by deep seismic reflection, Acta Geophys. Sinica 39(suppl.2), 29-35 (1995).

16. P. Tapponnier et al. Active thrusting and folding in the Qilian Shan, and decoupling between upper crust and mantle in northeastern Tibet, Earth Planet. Sci. Letter 97, 382-403 (1990).

17. W.J. Zhao, K.D. Nelson, and Poroject INDEPTH Team. Deep seismic reflection evidence for continental underthrusting beneath southern Tibet, Nature 336, 557-559 (1993).

18. Jung-sheng Tseng, Yung-chu Kan. Deep sub-basement reflection in the western part of Chaidamu Basin, Acta Geophys. Sinica 10(1), 54-66 (1996).

19. A. Hirn et al. Crustal structure and variability of the Himalayan border of Tibet, Nature 307, 23-25 (1984).

20. K.D. Nelson et al. INDEPTH and the fluid middle crust beneath Tibet, in "Global Tectonic Zones" Proc. 30th IGC (1996).

21. Rui Gao and Xiang-zhou Cheng. Preliminary geodynamic model of Golmud-Ejin Qi geoscience transect, Acta Geophys. Sinica 39 (suppl.2), 3-14 (1995).

22. R.K. Verma. Gravity field and nature of continent-continent collision along the Himalaya. Phys. and Chem. of the Earth 18 (1/2), 385-403 (1991).

23. M. Barazangi et al. Focal depths and fault plane solutions of earthquakes and active tectonics of the Himalaya, J, Geophys. Res. 89(B8), 6918-6928 (1984).

24. Jame Ni et al. Seismic images from the INDEPTH-2 broadband experiment: implications for underthrusting models, In "Global Tectonic Zones" Proc. 30th IGC (1996).

25. Jin Yu and M.K. Mcnutt. Mapping the descent of Indian and Eurasian plates beneath the Tibetan plateau from gravity anomalies, J. Geophys. Res. 101(B5), 11275-11290 (1996).

26. W. L. Zhao and W. J. Morgan. Injection of Indian crust into Tibetan lower crust: A two-dimensional finite-element study, Tectonics 6, 489-504 (1987).

27. Rui Feng and Hainan Zhou. Crustal Q-structure beneath the Tibetan Plateau, Acta Geophys.

Sinica 28(suppl.), 174-184 (1985).

28. Jianping Wu and R. Zeng. Inversion of Q-value structure beneath the Tibetan Plateau, Acta Seism. Sinica 9(2), 271-278 (1996).

29. N. Beghoul, M. Barazangi and B.L. Isacks. Lithospheric structure of Tibet and western North America: Mechanics of uplift and a comparative study, J. Geophys. Res. 98, 1997-2016 (1993).

30. S. D. Willet and C. Beaumount. Subduction of Asian Lithospheric mantle beneath Tibet inferred from models of continental collision, Nature 369, 642-645 (1994).

31. S. Ye et al. Crustal structure beneath the eastern Swiss Alps derived from seismic refraction data, Tectonophysics 242, 299-221 (1995).

32. Chun-yong Wang. A study on the deep structure of Dabieshan orogenic belt, Report in 30th IGC (1996).

Proc. 30 *Int'l. Geol. Congr.*, Vol. 6, pp. 133-140
Xiao Xuchang and Liu Hefu (Eds)
© VSP 1997

Comparative Studies of the Cenozoic and the Mesozoic Tectonics and Evolution between West Kunlun-Pamir (China) and Pyrenees (France)

QU GUOSHENG

Institute of Geology, SSB, Beijing, 100029, P. R. China

JOSEPH CANEROT

Lab. of Geology, Paul Sabatier University of Toulouse III, Toulouse, France

JIANG CHUNFA, WANG ZONGQI, ZHAO MIN

Institute of Geology, CAGS, P. R. China

Abstract

The Mesozoic and the Cenozoic intracontinental deformations and tectonics in West Kunlun-Pamir (China) and Pyrenees (France) show that: Pyrenees and West Kunlun-Pamir Mountains are part of disappeared Tethys and collided Himalayas in stage of Pyrenees-Himalayan. Arc of West Kunlun-Pamir shows the thrust (in Yecheng)-oblique thrust (in Zepu)-thrust (in Kashigar)-oblique thrust-thrust from the east to the west as similar as the Arc of Basque in the west of Pyrenees in France, and a series of trans-faults, pull-apart basin and earthquakes accompany with the arc active faults and arc tectonics during the Cenozoic intracontinental collisions. Foreland marine and continental basins develop in the north side of two mountains, that is, the North faulted belt of Pyrenees and Aquitain Basin, and the North faulted belts of West Kunlun-Pamir and Southwest Depression of Tarim Basin. These two mountains have the following stages of tectonic evolution: an intercontinental collision in the end of Hercynian, a rifting in the Jurassic - Cretaceous, and a compression stage since the Eocene. The displacements and uplifting by intracontinental collision in the Cenozoic in West Kunlun-Pamir Mountains are more larger than that in Pyrenees.

Keywords: Arc tectonics, West Kunlun-Pamir, Pyrenees, Tectonic evolution, Cenozoic intracontinental collisions

INTRODUCTION

The West Kunlun-Pamir and the Pyrenees are parts of Himalaya-Tethys where the rapid uplifting, thrusting, oblique thrusting and strike-slip displacement took place during the Cenozoic, especially the Quaternary. It is well known that Pyrenees is a typical intracontinental collided orogenic belts in Cenozoic in Europe and the world [8], and there are many geological and geophysical studies in Pyrenees since this century shown the tectonic divisions, nappes and arc tectonics, stratigraphy, metamorphic and magmatic basements and integrated interpretations of geological and geophysical cross-sections through the mountains for intracontinental orogenic models [3, 4, 5, 6, 7, 8, 12, 18, 19, 24, 25]. Comparative to Pyrenees, we got only few knowledge from West Kunlun-Pamir concerning with stratigraphy, metamorphism and magmatism, ophiolites, tectonic divisions, arc tectonics and deformation data [1, 2, 14, 15, 16, 20, 21, 22, 23]. By comparing West Kunlun-Pamir with Pyrenees, this paper will show the possible tectonics, the multi-orogenic and intracontinental collision model in the Cenozoic in West Kunlun-Pamir.

TECTONIC DIVISIONS

Tectonic Divisions in Py

Pyrenees, located between Iberie and Europe Plates, is a typical intracontinental collided orogenic belts based on Mesozoic rifting basins. Pyrenees have the following belts from north to south [7, 12, 25] (fig.1):

Fig. 1 Tectonic Division and Main Structures of Pyrenees Mountains (modify after Henry, 1987)

Outer or Small Pyrenees, located between Aquitain foreland basin and North Pyrenees, are the northern foreland faulted and folded belts of Pyrenees thrusting to the north with marine and continental deposits since the Mesozoic. The front thrusting faults of North Pyrenees are the south boundary of this belt [7, 25].

North Pyrenees are the domain thrusting systems and arc tectonics of intracontinental collided zone of Pyrenees composed by Mesozoic and Paleozoic marine sedimentary rocks, metamorphic rocks [3]. The front thrusting fault of Axis Pyrenees is the south boundary of this belt.

Axis Pyrenees are the root belts of large nappe systems from north to south [18] and the domain parts of Pyrenees with Hercynian metamorphic and granite basement rocks, The south boundary fault of this belt is the root zone of large Aragono Catalan thrust-detachment-thrust arc of South Pyrenees.

South Pyrenees are formed by large Aragono Catalan thrust-detachment-thrust arc where Mesozoic and Cenozoic marine and continental deposits thrust from north to south.

Tectonic Division in West Kunlun-Pamir

West Kunlun-Pamir, between Indian and Tarim Plates, are typical multi-collided orogenic belts since Paleozoic. The arc tectonics of West Kunlun-Pamir are formed in the Himalayan Orogeny [15, 20], and have the following belts from north to south (fig.2):

Faulted and Folded Belts of Southwest Tarim basin, located between Southwest Tarim Basin and West Kunlun-Pamir, is a foreland basin where the duplex and thrust nappes thrust from south to north. The boundary faults are the conceal active faults of Yingjisha-Hetian in the North, and the North front thrusting-oblique thrusting faults of West Kunlun-Pamir in the south [15].

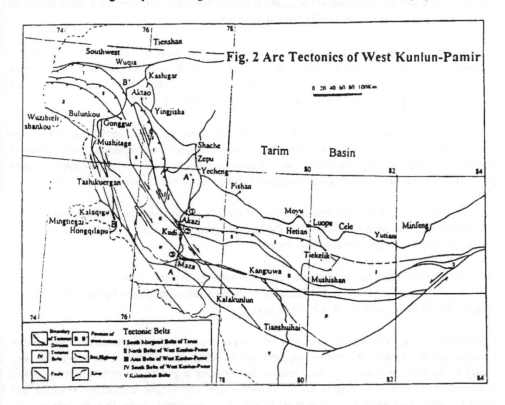

Fig. 2 Arc Tectonics of West Kunlun-Pamir

Arc and Thrusting Tectonics of Southwest Tarim is a part of Tarim basin which mixed to West Kunlun-Pamir by regional thrusting or oblique thrusting from north to south or West Kunlun-Pamir obduction (thrusting-oblique thrusting) to the north [26].

North Belts of West Kunlun-Pamir is the main Hercynian collision zone of paleo-plate tectonics [15, 21, 23], where some Carboniferous basic volcanic rocks thrust to the Jurassic continental deposits from south to north. The south boundary of this belt is the front thrusting fault of Axis West Kunlun-Pamir .

Axis West Kunlun-Pamir is the highest uplifting belts with Proterozoic, Paleozoic and Cenozoic metamorphic and granitic basements in West Kunlun-Pamir by large nappe systems from south to north in Cenozoic, for example, the root belts of Bulunkou-Aketashi nappe[10]. The south boundary of it is Mazha-Tashikuergan-south Kalakuli melange-mylonite suture zone in West Kunlun-Pamir which thrust to the north.

South Belts of West Kunlun-Pamir, consisted by Proterozoic, Paleozoic metamorphic rocks and Mesozoic marine deposits, also thrust from south to north.

Kalakunlun Belts locate in the south of Kalaqigu thrusting faults from south to north in Pamir and

DEFORMATION SINCE MESOZOIC AND CENOZOIC

Deformation in Pyrenees

Mesozoic: a rifting period in Pyrenees shown in the following[7]:

Triassic: Early Rifting. Continental carbonate and evaporate and then joint to a parts of Tethys based on the post Hercynian strike-slip and rifting structures.

Jurassic: Middle Rifting. Platform marine carbonate from the polar of Tethys to Atlantic by the northwest-southeast extension in early Jurassic (Lias). The thick calc-dolomitic carbonate deposits show the rapid NW-SE extension in middle Jurassic (Dogger). The limestone, mudstone and dolomite from the platform of Atlantic polar in late Jurassic (Malm).

Cretaceous: Generation of Pyrenees Rift. Pyrenees rift generated in the early Cretaceous, and became active continental margins in middle and late Cretaceous accompanying with pull-apart fylish basin, intensive diaprism and thinning of crust and middle grade metamorphism (Senonien (K_2)), and then the oblique convergence occurred between Iberie and Europe Plates[7, 12].

Cenozoic : arc tectonics by intracontinental collision. Many works show that the Pyrenees inverted thrusting and folding belts were formed by Pyrenees Orogeny in Cenozoic. Arc tectonics and nappe structures identify the typical tectonics in Pyrenees.

Pyrenees have two marginal arc tectonic belts: The Asturien, Basque, Lourdes, and Corbieres arc tectonics in the northern margin, and Cantabre, Aragono Catalan and Pedra Forca in the southern margin (fig.1) which can be divided into two types:

Collision-Thrusting Type (Basque Arc) in North Margin: In mapping, the kinematics of arcs change from the top (thrusting) to the transitional parts (oblique thrusting and trans-strike-slip faults), and finally to the end (thrusting). In cross-sections, the thrusting angles of the detachment faults and displacements of arcs change from the top (lowest angles and longest displacement) to the end (largest angles and shortest displacement), and the transitional parts (middle angles and displacement) (fig.3)[12].

Thrusting-Gravity Sliding and Detachment Type (Aragono Catalan) in South Margin: Gavarnie Nappes, the root belts of Aragono Catalan arc, consisting of basement unit, root or Gavarnie unit, Cotiella and Mont Perdu units, thrust along Gavarnie faults, and then sliding by gravity along the bottom lower angle detachment faults, and finally thrust along the front thrust faults of South Pyrenees belt with the pig-back thrust sequences from north to south in cross-sections [12, 18] and have the similar kinematics with the collision-thrusting type in mapping.

Deformation in West Kunlun-Pamir

Mesozoic: a rifting period in West Kunlun-Pamir shown in the following:

Triassic-Cretaceous: Marine basins develop in South and Kala-Kunlun-Pamir. The interplate subductions and collision took place along the Maza-Tashikuergan suture zone in the end of Triassic. The axis and north Kunlun-Pamir were uplifted and eroded during the Mesozoic.

Jurassic-Tertiary marine and continental deposits develop in the faulted and folded belts of southwest Tarim foreland basin, and arc thrusting tectonics of southwest Tarim. Field investigations and seismic profiles show that the southwest Tarim basin was marine and continental extension basins during Jurassic-Tertiary.

Cenozoic: arc tectonics by intracontinental collision and reorogeny. Many field investigations show that the West Kunlun-Pamir were formed and uplifted mainly by Himalayian orogeny during the Cenozoic.

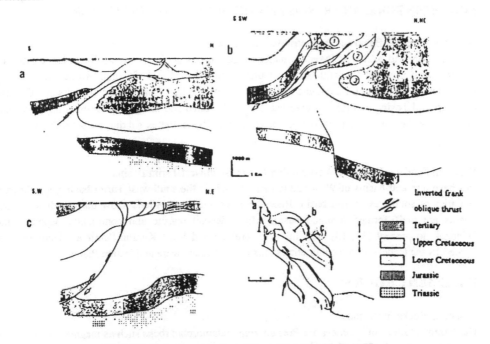

Fig. 3 Mapping and cross-sections of Basque Arc (after Henry, 1987)

Bulunkou Nappes: According to the field works, a large nappe of Axis Kunlun-Pamir along Bulukou-Gonggar root belts thrust over the late Paleozoic and Mesozoic rocks of arc thrusting tectonics of southwest Tarim and North Belts of West Kunlun-Pamir from south to north since Neogene [10, 20] (fig.4).

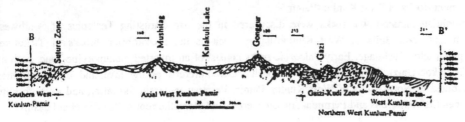

Fig. 4 Cross-section of China-Pakistan

Arc Tectonics in the Marginal Belts of West Kunlun-Pamir: In mapping, the kinematics of arc changes from the top (thrusting to the north in the south of Kashigar) to the transitional parts (oblique thrusting to the northeast and dextral trans-strike-slip faults in the south of Yingjisha-Zepu), and finally to the end (thrusting to the north in the south of Yecheng, oil field of Kekeya) (fig. 2), and a large Tashikuergan dextral pull-apart basin were formed since Neogene. In cross-sections, the thrusting angles of the detachment fault and displacement change from the top of the arc (lowest angles and longest displacement), the transitional parts (middle angles and displacement) to the end of the arc (largest angles and shortest displacement). Crossing the different parts of Zepu arc (one of small arc of West Kunlun-Pamir), the kinematic characteristics above also can be observed.

NORTHERN FORELAND BASINS AND THEIR TECTONIC EVOLUTION

Northern Foreland Basins of Pyrenees: Aquitain Basins

Northern foreland basins of Pyrenees include Aquitain Basins in the west, Carcassonne Basins in the east, and inverted foreland belts of Outer or Small Pyrenees. Seismic profiles and drillings show that Aquitain and Carcassonne Basins filled with the Mesozoic marine deposits and Cenozoic continental deposits, and formed inverted foreland faulted and folded basins thrusting from south to north by the Cenozoic intracontinental collisions, Iberie Plates subducted beneath the Europe Plate[7].

Northern Foreland Basins of West Kunlun-Pamir: Southwest Tarim Basins

Northern foreland basins of West Kunlun-Pamir include the southwest Tarim basins, arc thrusting tectonics of southwest Tarim basin. Recent studies and explorations show that southwest Tarim Basins are multi-formed foreland basins where deposit marine and continental deposits since Paleozoic to present [11, 13]. The inverted foreland and West Kunlun-Pamir arc tectonics were formed by the Cenozoic intracontinental collisions between Tarim and Indian Plates.

BASEMENT ROCKS

Basement Rocks In Pyrenees

The basement rocks of Pyrenees are Precambrian metamorphic rocks such as magmatite, gneiss and gneissic granites distributed in North Pyrenees Belts, the Basque block, and east Pyrenees [19]. Ordovician and Silurian lower metamorphic marine rocks, Devonian and Carboniferous marine limestone were located in the Axis Belts, North Belts of Pyrenees, Mouthoumet and Black Mountains of France [9]. A large unconformity between Carboniferous and Permian-Triassic and Permian molassic formation in the Axis Belts of Pyrenees represent the Hercynian orogeny.

Basement Rocks in West Kunlun-Pamir

Precambrian metamorphic rocks were discovered in the Arc Thrusting Tectonics of Southwest Tarim and Axis Belts of West Kunlun-Pamir, such as the Precambrian blocks of Tiekelike, Gongguar and Mushitage. Early Paleozoic rocks consist of an inverted metamorphic zone, such as in Bulunkou-Gaizi. Devonian marine deposits, Carboniferous pillow lava and basic volcanics distributed in North Belts of West Kunlun-Pamir. Permian molassic formation and unconformity between Carboniferous and Permian-Triassic were found along the root belts of foreland basins.

TECTONIC EVOLUTION

Hercynian Orogeny

West Kunlun-Pamir: Kudi-Wuyitake ophiolite zone represent the collision zone between the Axis Belts of West Kunlun-Pamir and Tarim Basins by Hercynian orogeny [14, 16, 17, 21, 23]. Intensive deformation took place in the end of Carboniferous.

Pyrenees: Basement rocks of Pyrenees and Europe Plates were formed by Hercynian orogeny and the Paleozoic marine rocks were faulted and folded very intensively in the Axis Belts of Pyrenees, Mouthoumet and Black Mountains of France.

Mazha-Tashikuergan melange and mylonitic suture zones show the collision between the Axis and South Belts of West Kunlun-Pamir and the Paleo-Tethys is disappeared. Some intermountains basins develop in Pyrenees.

Mesozoic rifts or extension
The rifting or extension basins were formed during the Mesozoic in both mountains.

Himalayan and Pyrenees orogenies
Inverted tectonics and foreland basins were formed by intracontinental collision orogenies in both mountains.

COMPARATIVE STUDIES AND CONCLUSIONS

Kinematics of thrusting-colliding type of arc tectonics in both mountains are similar: thrust-oblique thrust-thrust-oblique thrust-thrust from the east to west. All the core of arcs is composed of the basement blocks. A series of dextral strike-slip faults in the east trans-margin and sinistral strike-slip faults in the west trans-margin of arc, pull-apart basin, active faults and earthquakes were developed during the formation of arc tectonics in Cenozoic.

West Kunlun-Pamir and Pyrenees are parts of disappeared Tethys and collided Himalayas. As Pyrenees, West Kunlun-Pamir are the intracontinental orogenic belts caused by the Cenozoic intracontinental thrusting and uplifting. The deformation degree and uplifting scale of intracontinental collision in West Kunlun-Pamir Mountains are stronger than that in Pyrenees.

The rifting or extension basins with the marine and continental deposits were developed in both mountains in the Jurassic and the Cretaceous.

Foreland basins develop in north side of both mountains, the North Pyrenees, small Pyrenees and Aquitain Basin, the arc tectonics of West Kunlun-Pamir and foreland southwest Tarim Basins. The Mesozoic rifting or extension basins changed to inverted foreland basins in the Cenozoic.

West Kunlun-Pamir are the Multi-collisions orogenic belts: intercontinental collisions in Hercynian and intracontinental collision in the Cenozoic. Pamir arc tectonics were only formed in the Cenozoic.

Acknowledgments

Thanks a lot for Prof. J. Rey, C. Majeste-Menjoulas, Delvolve , Lab. de Geologie, Universite Paul Sabatier for their helpful supports in the field investigations in Pyrenees, and Prof. Chen Jianqiang, Department of Geology and Mine, Chinese Geological University for the helps of field works in West Kunlun-Pamir.

REFERENCES

1. Bureau of Geology and Mineral Resources of Xinjiang Uygur Autonomous Region. Geological map of Xinjiang Autonomous Region (1:2000000). Geological Publishing House (1985).

2. Bureau of Geology and Mineral Resources of Xinjiang Uygur Autonomous Region. Regional Geology of Xinjiang Uygur Autonomous Region. Geological Publishing House (1993).

3. J. Canerot, et E-J. Debroas. Remise en question du "Complexe de resedimentation albo-cenomaien" de la zone nord-
Pyreneene. C. R. Acad. Sci. Paris. 307. Serie II. 1711-1716 (1988)

4. J. Canerot, et J. L. Lenoble. Le diapir du Lichancumendy (Pyrenees-Atlantiques), nouvel element de la marge iberique des Pyrenees occidentales. *C. R. Acad. Sci. Paris*, 308, Serie II, 1467-1472 (1989).

5. J. Canerot. Comparative study of the Eastern Iberides (Spain) and the Western Pyrenees (France) Mesozoic basins. Palaeogeography, *Palaeoclimatology, Palaeoecology*, 87, 1-28 (1991).

6. J. Canerot, et J-L. Lenoble. Diapirisme cretace sur la marge iberique des Pyrenees occidentales: exemple du pic de Lauridle; comparation avec l'Aquitaine, les Pyrenees centrales et orientales. *Bull. Soc. geol. France*, 164, 719-776 (1993).

7. J. Canerot, et C. Majeste-Menjoulas. Excursion geologique dans la chaine Pyreneenne et ses bassins d'avant-pays nord et sud. *elf aquitaine production* (1994).

8. M. Daignieres, J. Canerot, B. Damotte, E.-J. Debroas, P. Desegaulx, J. Estival, A. Hirn, M. Levot, M. Sequret, et A. Villien. Profil ECORS Arzacq-Pyrenees ouest. *Rapport d'implantation* (1990).

9. Jean-Jacques Delvolve, Pierre Souquet, Daniel Vachard, Marie-France Perret et Pascale Aguirre. Caracterisation d'un bassin d'avant-pays dans le Carbonifere des Pyrenees: facies, chronologie de la tectonique synsedimentaire. *C. R. Acad. Si. Paris*, 316, Serie II, 959-966 (1993).

10. Peng Gao et al. Tectonic characters of thrust nappes in Aketashi area of West Kunlun. *Xinjiang Geology*, 12, 115-123 (1984).

11. Dengfa He, Xiuxiang Lu, Yonghan Lin, and Dazhong Dong. Foreland basin analysis. Petroleum Industry Publishing House (1996).

12. J. Henry. Enquete geologique sur les Pyrenees "Mes cinq dernieres minutes". *Bulletin Technique Exploration Production. Societe national elf aquitaine* (1987).

13.. Chengzao Jia, Guoqi Wei, Huijun Yao and Liangchen Li. Tectonic evolution and regional structural geology. *Book series on petroleum in the Tarim basin*. Petroleum Industry Publishing House (1995).

14. Chunfa Jiang, Jingsui Yang. The geological and tectonic outlines of Kunlun Mountains. *Journal of Institute of Geology*, CAGS, 15, 70-80 (1986).

15. Chunfa Jiang, Jingsui Yang, Binggui Feng. Kunlun opening and closing tectonics. Geological Publish House (1992).

16. Chunfa Jiang. Soft and solid basements and basement suture zone. *Chinese Geology*. 3, (1993).

17. Shenglui Jiao. Plate tectonic evolution of northwest margin of Tibet Plateau. *Geological Papers on Tibet Plateau (1)*, 164-174 (1982).

18. C. Majeste-Menjoulas. Evolution Alpine d'un segment de chaine varisque: nappe de Gavarnie, chevauchement cinq-monts-gentiane (Pyrenees centrale et occidentales). *Travaux du Laboratoire de Geologie-Petrologie et du Laboratoire de Geologie mediterraneene et Pyreneenne Associe au C. N. R. S.*. These doctoral de Universite Paul Sabatier Toulouse (1979).

19. C. Majeste-Menjoulas, and P. Debat. Pyrenees. J. D. Keppie (Ed.): Pre-Mesozoic Geology in France and Related Area (1994).

20. Guosheng Qu, Joseph Canerot, Zongqi Wang, Min Zhao. Arc tectonics in orogenic belts. Acta Geologica Sinica, 70 (1996).

21. Yusheng Pan. Tectonic characters and evolution of West Kunlun Mountains. Acta Geologica Sinica, 64, 224-231 (1990).

22. Yusheng Pan. Integrated introduction on scientific researches of Kalakunlun-Kunlun Mountains. Meterological Publishing House (1992).

23. Yusheng Pan, Yi Wang, Ph. Matte, P. Tappornier. Tectonic evolution along the geotravers from Yecheng to Siquanhe. Acta Geologica Sinica, 68, 295-307 (1994).

24. M. Sequret. Etude tectonique des Nappes et series decollees de la partie centrale du versant sud des Pyrenees (caractere synsedimentaire, role de la compression et de la gravite). These Sc., Mpntpellier et Pub. USTELA, Momtpellier (1970).

25. P. Souquet, B. Peyberne, M. Bilotte, E.-J. Debroas, J. Rey, et J. Canerot. Nouvelle esquisse structurale des Pyrenees. Travail realise par le Laboratoire de Geologie de l'Universite Paul Sabatier, Toulouse (1977).

26. Suyun Wang, Zhenliang Shi, Wenlin Huan. A destroyed earthquake with middle depth of epicenter in Yecheng, Xinjiang in 14, Feb. 1980. *Acta Seismologica Sinica*, 14, 137-143 (1992).

Proc. 30ᵗʰ Int'l. Geol. Congr., Vol. 6, pp. 141-153
Xiao Xuchang and Liu Hefu (Eds)
© VSP 1997

Lithospheric Fabric of the Qinghai – Tibet Plateau

CUI JUN-WEN

Institute of Geology,Chinese Academy of Geological Sciences,Beijing   100037, China

Abstract

The Tethyan-Himalayan domain is a gigantic orogenic complex of "convergence-intracontinental subduction type" , which is called the Tethyan-Himalayan orogenic complex. Based on the lithospheric texture, structure and present dynamic system, three types of lithospheric fabric are recognized: the Himalayan thrust – superposition type,the Qinghai-Tibet strike-slip – stretching type and the Kunlun-Altun-Qilian thrust – spreading type. The deep processes in the complex revealed by the characteristics of the Cenozoic deformation and lithospheric fabric are: bidirectional subduction and N-S asymmetrical spreading in the foreland and backland; and deep-level thermal-uplift stretching and E-W asymmetrical spreading in the hinterland. According-ing to their formation mechanism and orogenesis, the mountain chains of the complex can be divided into three genetic types: Himalayan type,Gangdise type and Kunlun (or Qilian) type.

Keywords: Tethyan-Himalayan orogenic complex, foreland, backland, hinterland, lithospheric fabric, stretch-ing(spreading,extension), Himalayan type, Gangdise type, Qilian (or Kunlun) type

INTRODUCTION

The Qinghai-Tibet plateau, which has long been known as the "third pole" of the earth, is a gigantic orogenic complex[5,19]of convergence-intracontinental subduction type formed by multiple split-ting, convergence, intracontinental subduction and amalgamation of the following 7 terranes[2,16]: the Altun-Qilian and North Kunlun-Qaidam terranes of the Tarim-Sino-Korean plate; the South Kunlun and Hoh Xil-Bayan Har terranes of the South China-Southeast Asia plate;and the Qiang-tang, Gangdise and Himalayan terranes of the Gondwana.Under the effect of Cenozoic strong deformation[2,14,19]and abrupt uplifting[2,3,6,19]it has become the earth's largest flexible massif with the thickest crust. On its outside are the India, Tarim, Alxa-Ordos and Yangtze rigid massifs or cratons separated from one another by piedmont sag zones.The plateau is divided,based on the characteristics of the Cenozoic deformation and lithospheric fabric,into 3 first-order tectonic units (domains or provinces)and 9 second-order tectonic units (belts or zones), which are separated by large fault (Fig.1). The three first-order units are the Himalayan foreland, the Qinghai-Tibet hinter-land and the Kunlun-Altun-Qilian backland.

LITHOSPHERIC FABRIC OF HIMALAYAN THRUST – SUPERPOSITION TYPE

The Himalayan thrust – superposition domain is composed of 3 second-order tectonic units:the North-Himalayan gliding-nappe belt, Himalayan foreland thrust-overlap belt and the Siwalik fore-land sag belt (Fig.1). The second belt is predominant. It consists of a series of south-thrusting faults or ductile nappe shear zones, from north to south, including the Yarlung Zangbo fault, Lhagoi Kangri-Kangmar thrust , North Himalayan ductile nappe shear zones[1,16], the Main Central thrust, the Main Boundary thrust, the Main Front thrust and the Main Siwalik thrust, which were started respectively in the Late Cretaceous-Early Eocene, Late Eocene, Late Eocene-Early Miocene,Early

142

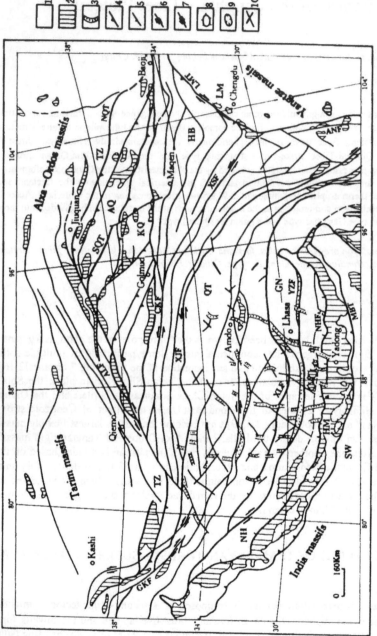

Figure 1, Schematic tectonic map showing the Tethyan-Himalayan orogenic complex. 1,sedimentary cover; 2,crystalline basement; 3,fault depressional zone; 4,thrust; 5,normal fault; 6,dextral strike-slip fault; 7,sinistral strike-slip fault; 8,metamorphic core complex or nappe; 9,gliding-nappe; 10,conjugate strike-slip faults. MBT, Main Boundary thrust; NHF, North Himalayan normal fault; YZF, Yarlung Zangbo fault; XLF, Xainza-Lhari normal fault;XJF,Xijir Ulan-Jinsha fault; CKF, Central Kunlun fault; SQT,South Qilian thrust; NQT, North Qilian thrust; GKF, Karakorum fault; XSF, Xianshui He fault; ALF,Altun fault; LMT, Longmenshan fault. Tectonic units (from south to north): foreland, SW, Siwalik foreland sag belt; HM,Himalayan foreland thrust-overlap belt; NH,North Himalayan gliding-nappe belt; hinterland, GN, Gangdise-Nyainqentanglha strike-slip – thermal-uplift extensional belt; QT,Qiangtang strike-slip – thermal-uplift spreading belt; HB, Hoh Xil-Bayan Har strike-slip – thermal-uplift spreading belt. TZ, South Tarim-Zoulang backland sag belt. backland, KQ, Kunlun-Qaidam thrust-spreading belt; AQ, Altun-Qilian backland thrust-extensional belt; TZ, South Tarim-Zoulang backland sag belt.

Miocene, Late Pliocene, Late Pleistocene and Holocene[2,7,19].The faults as a whole spread in the direction consistent with that of their thrusting,forming a south-spreading progressive thrust sequance.In section these faults assume a piggyback pattern[1,5], extending downward to converge at the intracrustal detachment layer (ICD) (Fig.2).The ICD is a dominantly granitic mylonite layer[2]between the upper and lower crust with the Vp=5.6-5.7km/s, the electrical resistivity being only 10^0-$10^1 \Omega$m and the average thickness of about3-6km. It dips about 10° north[11].At Kangmar the ICD occurs at the depth o f about 35 km[16].The North Himalayan ductile mylonite zone represents an outcrop of the ICD at the surface. The mylonite,formed at 580−680℃, is metamorphosed at a high-temperature amphibolite facies[2]. The Siwalik foreland sag belt is a syntectonic basin controlled by both the Himalayan progressive thrusting and the strong subduction of the India plate beneath the Himalaya. It is basically synchronous in origin with the North Himalayan ductile nappe shear zone and the Main Central thrust [6].The intense deformation[6] of the 6000-m-thick Middle Miocene-Pleistocene molasse in the sag belt,especially the development of a series of thrusts such as MFT and MST,suggests that the sag belt is the southern dynamic boundary of the Tethyan-Himalayan orogenic complex. Different in genetic mechanism from the sag belt,the North Himalaya has the nature of a post-orogenic tectonic basin of gliding-nappe type,in which the sedimentary sequence contains a series of gently-dipping E-W lag normal faults or brittle-ductile gliding-nappe shear zones.The gliding-nappe shear zone at the bottom of the sedimentary is termed the basement detachment layer(BSD). The North Himalayan brittle-ductile mylonite zone is abasement detachment layer exposed at the surface, which has a formational temperature lower than 350° and is metamorphosed at a lower greenschist facies [2].The extensive development of normal slide-type flow cleavages and NEE and E-W stretch lineations[2] in the BSD indicates a gravitational gliding-napping of the Tethyan sedimentary sequence along the BSD from the periphery to the interior.The gliding-napping began in the Middle Miocene-Pliocene (about 20-3Ma) when the MCT and MBT were formed, and became intense since the Latest Pliocene(3Ma)when the Qinghai-Tibet plateau was abruptly uplifted[6]. The lag normal fault system in the Himalayan thrust-superposition domain gradually spread southward with the progressive thrusting. The transition at the late stage from the ductile nappe shear zone to a normal slide shear zone for the North Himalaya[2,7]implies that the south boundary of the gliding-napping was extended into the upper part of the High Himalayan crystalline sequence.

Deep structural activity in the Himalayan thrust – superposition domain is manifested by an over-step-type (or retreat-type) northward subduction[2,5,6], which is in contrast to the southward progressive thrusting in the upper continental crust and results in the formation of a subduction-accretionary wedge at th e bottom of the lower continental crust[1]and the offset of the Moho[10]. The most remarkable subduction wedge corresponds to the Lhagoi Kangri-Kangmar island chain-like metamorphic core complex zone (Fig.2), is underlain by a low-susceptibility zone and disturbed seismic velocity zone formed by thermal anomaly or local magmatism, and has surficial heat flow values as high as 91-146mW/m^2[4,16].These suggest that the site of the wedge is most likely the front of the present subduction of the India plate.The geometry of the wedge and the local thickening of the lower continental crust indicate an intense shearing between the crust and mantle.Therefore the Moho "layer"is structurally considered as a deep ductile shear zone with a four-dimensional extension(i.e.the extent and thickness of the "layer"vary with time). This "layer"is called the crust-mantle detachment layer (CMD)[2,3,5]. In section,the Himalayan thrust-suporposition domain is in the form of a wedge shortened in the deep part and expanded in the shallow part ,while the Moho takes the shape of a basin with the center at the Kangmar subduction wedge and the two ends tilting upward. The crust thickness is 70km beneath the Yarlung Zangbo River area, 55 km southward at the India platform and up to 80km[16]in the Kangmar area. Gravity data show that the Moho beneath the Himalayan thrust – superposition domain dips 20° north[11].

LITHOSPHERIC FABRIC OF QINGHAI-TIBET STRIKE-SLIP-STRETCH TYPE

144

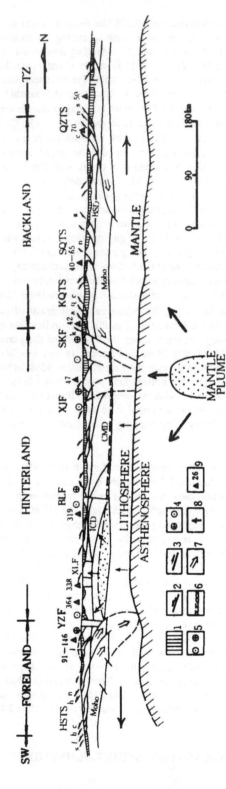

Figure 2, Geoscience trasect crossing the Tethyan-Himalayan Orogenic Complex. 1,sedimentary cover; 2, thrust; 3,normal fault; 4,dextral strike-slip fault; 5,sinistral strike-slip fault; 6,mixed crust-mantle layer; 7,underthrusting direction; 8,direction of thermal-uplift; 9, heat flow (mW/m²). BSD, basement detachment layer; ICD, intracrust detachment layer; CMD,crust-mantle detachment layer. HSTS, Himalayan-Siwalik thrust system:s,Main Siwalik thrust (MST),f,Main Frontal thrust (MFT), b,Main Boundary thrust (MBT),c,Main Central thrust (MCT),h, North Himalayan thrust (NHT); n, North Himalayan normal fault (NHF); l,Lhagoi kangri-Kangmar thrust (LKT); YZF,Yarlung Zangbo fault; XLF,Xainza-Lhari normal fault; BLF, Bangong-Lancang fault; XJF,Xijir Ulan-Jinsha fault; SKF,South Kunlun fault; KQTS,North Kunlun-Qaidam South Marginal thrust system:k,Central Kunlun fault (CKF); a, Arkatag thrust (ART); q, Qimantag thrust (QMT); c,Qaidam Central thrust (QCT); SQTS,South Qilian-Qaidam North Marginal thrust system : s,South Qilian thrust(SQT): n,Qaidam North Marginal thrust (QNT); QZTS, North Qilian-Zoulang South Marginal thrust system; c,Central Qilian thrust system; c,Central Qilian thrust (CQT); n,North Qilian thrust (NQT); s,Zoulang South Marginal thrust (ZST). SW,Siwalik foreland sag belt; TZ,South Tarim-Zoulang backland sag belt.

The Qinghai-Tibet strike-slip – stretch domain is composed of two second-order tectonic units: the Northern Tibet strike-slip – thermal-uplift extensional beltand the Hoh Xil-Bayan Har strike-slip – thermal-uplift spreading belt.The former can be recognized the Gangdise-Nyainqentanglha strike-slip – thermal-uplift extensional belt and the Qiangtang strike-slip – thermal-uplift spreading belt.The domain is tectonically basically characterized by intense nonuniform deformations of strike-slip, unified vertical extension and E-W spreading.

The Qinghai-Tibet domain is a part of the Tethyan-Himalayan orogenic complex that has the thickest crust and relatively thin lithosphere.Here the crust is 75-80km in thickness,the upper and lower crust are 23-29 km and 31-50 km respectively, and the Moho morphology is stable. At the bottom of the continental crust there is generally a mixed crust-mantle layer with the Vp of 6.7-7.4 km/s(Fig.2). The layer is the thickest in the Gangdise area,being 26-29 km, and decreases northward to 5 km[16].Magnetotelluric data suggest that the lithosphere is 100-130km thick[16,21]. The Gangdise area is the area with the highest heat flow(up to 338-3 64 mW/m^2 [16])in the complex,and the maximum temperature measured in 2006-m-deep well is 262.03℃.Northward from the area the heat flow gradually decreases, dropping to the lowest(47mW/m^2)in the hinterland.Natural seismic converted wave study suggests that in the Hoh Xil-Bayan Har-Qiangtang area the lower crust-upper mantle exhibits a multi-layer lithospheric texture of alternating high- and low-velocity zones, and there is a lung-shaped low-velocity anomalous body at the depth of 200-360km in Hoh Xil[20]. This is inferrred to be related to the partial melting of the continental crust and lithospheric mantle caused by multiple strong subductions of the Tarim plate benenth the Kunlun Mountains since the Late Cretaceous[6]. Based on the temperature of 1000℃ at the bottom of the continental crust [16], the intracrustal geothermal fields since 65-40Ma, 40-20Ma, 20-3Ma and 3Ma have been simulated using the multiple subduction models of geothermics, elastic - plasticity and extensive deformation. The simulation indicates that the Gangdise area is the area with the greatest geothermal growth rate,up to 137.5℃/Ma since 3Ma, and displays upward-arching isothermal contours. In contrast,the isothermal contours in the Hoh Xil-Bayan Har area assume an inharmonious pattern slightly curving down in the upper part and arching a little in the lower[6].

The Qinghai-Tibet hinterland is characterized by the development of high-angle longitudinal normal strike-slip faults and low-angle thrusts and normal faults (Figs,1,2). Stretching lineations show that the sedimentary cover of the hinterland has experienced a gravitational gliding-nappe from the periphery to the center which is similar to that of the North Himlayan belt. For instance, the Gangdise and Kunlun areas generally lack a sedimentary cover due to uplift of the crystalline basement, whereas the Bayan Har area, located at the center of the gliding nappe, has a sedimentary cover up to 25 km thick. The sedimentary cover of the Bayan Har area is the thichest in the Tethyan-Himalayan orogenic complex (Fig.2)and tends to be thicker and thicker eastward[6]. The Xainza-Lhari normal faullt[1]is a basement detachment layer exposed at the surface(Fig.2). The high-angle faults in the Qinghai-Tibet hinterland have different kinematic characteristics. Northern Tibet is dominated by conjugate NE and NW strike-slip, E-W normal strike-slip and N-S fault depression. Large normal strike-slip faults, such as the Xijir Ulan-Jinsha fault, Bangong-Lancang fault and Xainza-Lhari fault, show at the surface the geometric features of conjugate strike-slip faults(Fig.1). This implies that the kinematics of the lithosphere beneath Northern Tibet is manifested by thermal-uplift extension (vertical) and E-W spreading(horizontal). The E-W spreading was mainly a tectonic event since 3Ma. The Paleocene-Eocene and Pliocene intense volcanic eruptions and magmatic effusions in Northern Tibet were related to local tension caused by thermal uplift extension which started in the Paleocene-Eocene[6]. A series of WNWto E-Wsinistral strike-slip faults and their intercalated longitudinal striped fault slices(Fig.1)exposed at the surface in the Hoh Xil-Bayan Har area suggest that the lithospheric kinematics in the area is dominated by successively east-squeezing lateral flow (eastward spreading). In the west segment(Hoh Xil), the intense sinistral strike-slip has led to the eastward flow of material, the streching thinning and necking of rock blocks, the formation of NE tensional and pull-apart basins, and the strong eruption of rift-type alkaline basalt since the Pliocene. In the east segment (Bayan Har), which also underwentwestward compression due to the northwestward subduction of the Yangtze plate,there are a series of south-protruding arcuate thrusts and bidirectional compressive folds in the Songpan ditrict[18]. The east limb of the arcuate thrust system is juxtaposed at low angle with the Longmenshan dextral strike-slip fault (Fig.1). The rock blocks on the north side of the system were thrust sorthward one by one, forming a thrust-superposition body in the upper continental crust. All this implies that in the upper continental crust of the Songpan massif, a second-order southward flow (spreading)has been super-posed on the regional dynamic background of eastward flow (spreading).

LITHOSPHERIC FABRIC OF KUNLUN-ALTUN-QILIAN THRUST – SPREADING TYPE

The Kunlun-Altun-Qilian thrust – spreading domain consists of 3 second-order tectonic units: the Kunlun-Qaidam thrust – spreading belt, the Altun-Qilian backland thrust – extensional belt and the South Tarim-Zoulang backland sag belt(Fig.1). The domain is basically characterized by a basin-range tectonic system consisting of south-dipping WNW arcuate thrusts or ductile nappe shear zones and their intercalated sag-type and lag-extensional basins [6]. Major thrusts or ductile nappe shear zones are,from north to south, as follows: the North Qilian thrust, Central Qilian thrust, South Qilian thrust, Qaidam North Marginal thrust, Qaidam Central thrust, Qimantag thrust and Arkatag thrust. These thrusts or shear zones on the whole dip south, getting increasingly young from north to south[6]. The fault system is opposite in propagation direction to the thrusting but is corresponding in composition to the Himalayan foreland progressive thrust sequence (Fig.2). In section the fault system shows an overstep propagation,i.e., later faults are suporposed on earlier ones. Three retreat-type thrust sequences are distinguished based on deep structural features. They are,from north to south,a follows: the North Qilian-Zoulang South Marginal, the South Qil-ian-Qaidam North Marginal and the North Kunlun-Qaidam South Marginal(Fig.2), which fall respectively on three gravity gradient zones at the north margin of the Qinghai-Tibet plateau[12]. In particular, the first thrust sequence coincides with the first-order gravity gradient zone of the Tethyan Himalayan orogenic complex[6]. The trunk faults of the three thrust sequences are respec-tively the Central Qilian thrust, the Qaidam North Marginal thrust and the Central Kunlun fault(Fig.2). These trunk faults are listric in section. In the overlying rock series are reversely-dipping thrusts or ductile nappe shear zones that form a fan-shaped structure superimposed on the retreat type thrust sequence(Fig.2). In North Qilian, Qaidam North Marginalarea and Kunlun at the center of the fan shaped structure, there are exposed deep-source rocks such as Early Paleozoic ophiolitic melange, high-pressure metamorphic rocks,eclogite and lower-crust crystalline rocks which are related to tectonic extension(diapirism). The North Qilian-Zoulang South Marginal and South Qilian-Qaidam North Marginal retreat-type thrust sequences were formed in the Triassic-Quaternary, while the North Kunlun-Qaidam South Marginal retreat-type thrust sequence was produced after the Oligocene[7]. The formation of the Qaidam massif had a bearing on the intense thrust-nappe processes (northward spreading)of the South Qilian-Qaidam North Marginal retreat-type thrust since the Triassic. In the Cenozoic,especially the Miocene-Pliocene, an abrupt depres-sion took place and a Cenozoic continental sedimentary formation up to 10,000m thick was accu-mulated in the depressed area[6]. The South Tarim-Zoulang sag belt in the backland of the oro-genic complex, which is corresponding to the Siwalik sag belt in the foreland, is the result of a combined effect of the southward subduction of the Tarim-Alxa massif and the northeastward oblique thrusting of the West Kunlun-Qilian mountain system. From the east to the west the South Tarim-Zoulang sag belt becomes increasingly young in age and progressively strong in depress-ing[6]. The molasse, formed since the Cenozoic, especially the Pliocene,is up to 10,000m thick. Reflection seismic survey data suggest that two sets of reflectors,dipping gently south and north respectively, exist in the upper continental crust of the sag belt [17]. The presence of vertical flow cleavages and recumbent folds with a slightly north-dipping axial plane in the crystalline basement of the Pre-Changcheng System[6] indicates that the southward subduction of the Tarim-Alxa massif was accompanied by an upward vertical structural extension (diapirism).

Within the Kunlun-Altun-Qilian tectonic domain,the crust is 40 – 70km thick and the average seismic velocity in the continental crust is 6.10 - 6.35 km/s. The thicknesses of the upper and lower crusts are 20 - 26km and 14 - 35km, respectively. At the bottom of the upper crust there is commonly a low-velocity layer which is about 6 - 10km in thickness and has a seismic velocity of 5.67 - 6.08 km/s[8]. The velocity tends to increase northward from 5.67 km/s in the Kunlun Mountains to 6.08 km/s in the Qilian Mountains. The Moho on the whole is gradually dipping southward. It is at the depth of only 40km in the Zoulang sag belt but at 63 km in the East Kunlun mountains. South of the Har Lake, it is offset. In the Central and South Qilian area on the north side, the Moho is buried at the depth of 60-70km [8]. Here the backland crust attains its maximum thickness. The Moho is uplifted in the Qaidam basin on the south side (Fig.2). This is associated with the intense intracontinental subduction of the Tarim-Alxa massif along the North Qilian-Zoulang South Marginal fault system [6]. Magnetotelluric sounding suggests that the lithosphere is 145-155km thick and the lithosphere-asthenosphere interface is slightly wavy in form, while the Qaidam North Marginal tectonic belt and the Zoulang sag belt are somewhat uplifted [22]. The heat flow values in the backland areas are comparatively low: 40-65mW/m^2 in the Kunlun-Qaidam thrust spreading area, 70mW/m^2 in the Altun-Qilian backland thrust-extension area and 50mW/m^2 in the Zoulang sag belt[13].

DEEP PROCESS IN THE TETHYAN-HIMALAYAN OROGENIC COMPLEX

The remarkable inhomogeneity of the lithospheric fabric of the orogenic complex reflects a complex dynamic mechanism and deep process during the formation and evolution of the mountains.

Foreland and Backland: Bidirectional Subduction and N-S Asymmetric Spreading.
After the closure of the Neo-Tethys the India plate was continuously and rapidly subducted beneath the Eurasia plate. Based on paleomagnetic data it is estimated that the subduction amount of the continental crust of the two is 2076-2681km [2,4,9,16] and the average subduction rate is 3.19-4.12cm/a. Correspoudingly the Himalayas was thrust or overthrust southward (spreading). Measurement of modern geo-deformation shows that the northward movement rate of the India plate is 4-6cm/a, the southward movement rate of the Himalayas is 1- 2cm/a, and the converging rate between the two is 5.08cm/a[15]. The India plate was subducted beneath the Himalayas at the angle of 20-30° along the intracrustal detachment layer and, after passing through the Lhagoi Kangri-Kangmar belt, penetrated into the lithospheric mantle at a high angle(Fig.3). Natural seismic data suggest that its subduction is down to the upper mantle. The arcuate subduction zone composed of the intracrustal detachment layer and Lhagoi Kangri-Kangmar fault forms the south marginal spreading (dynamic) boundary of the orogenic complex (Fig.3). With the continuous, intense northward subduction of the India plate, the continental crust of the Himalayan terrane was abruptly shortened, whereas the lithospheric materials migrated upward along the subduction zone and constantly spread and moved southward by means of progressive thrusting or overthrusting [1,5,7]. The formation of the Lhagoi Kangri-Kangmar island chain-like metamorphic core complex zone is related to the upward migration of the lithospheric materials caused by the intense subduction of the India plate and to the resultant vertical extension. In contrast, the High Himalayan crystalline sequence represents a lower crust component exposed at the surface due to the upward spreading of material from depth. The southward spreading of the Himalayan thrust - superposition domain began in the terminal Late Cretaceous Eocene (65 - 40 Ma).The Qinghai-Tibet plateau rapidly uplifted in the Oligocene-Latest Pliocene (40-3Ma)and entered a fast uplift stage or intense spreading period since the Latest Pliocene(3Ma). The estimated average spreading rate since 65Ma is 2.00 - 3.25mm/a, much lower than 1-2cm/a [6], the present southward motion rate for the Tibet plate obtained by geo-deformation measurement.This indicates that the southward spreading of the Himalayas continuously speeded up with increase of the uplifting rate.

The Kunlun-Altun-Qilian thrust - spreading domain is dynamically characterized by dual subduction and spreading rotation. This means that its peripheral rigid block was subducted southward

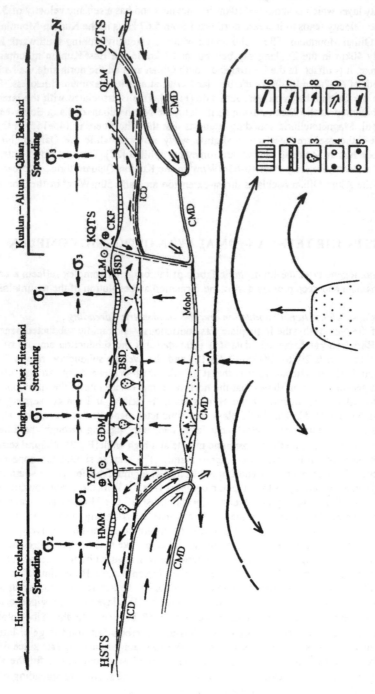

Figure 3. Sketch showing dynamic processes of uplifting of Tethyan-Himalayan orogenic complex . 1,sedimentary cover; 2,partially molten layer; 3,magma pocket or mantle plume; 4,dextral strike-slip fault; 5,sinistral strike-slip fault; 6,underthrusting boundary; 7,spreading boundary; 8,substance flows direction; 9,underthrusting direction;10,spreading direction; BSD,basement detachment layer; ICD,intracrustal detachment layer; CMD,crustmantle detachment layer; L-A, lithosphere-asthenosphere interface; YZF,Yarlung Zangbo fault; CKF, Central Kunlun fault; HSTS, Himalayan-Siwalik thrust system; KQTS, North Kunlun-Qaidam South Marginal thrust system; QZTS, North Qilian- Zoulang South Marginal thrust system. HMM, Himalayas; GDM,Gangdise Mts; KLM, Kunlun Mts; QLM, Qilian Mts.

along the North Qilian-Zoulang South Marginal fault system and the North Kunlun-Qaidam South Marginal fault system(Fig.3),whereas the Kunlun and Altun-Qilian massif was obliquely thrust northeastward. As a result, its backland part, while spreading northward, rotated clockwise at 18-30° relative to the peripheral rigid block (Fig.4).The northward intense spreading (thrusting and overthrusting) and rotation in the backland area has resulted in the formation of the Altun sinistral strike-slip fault[6]. Deep-source rocks, such as crystalline series,ophiolitic melange and high-pressure metamorphics, are widely exposed in the spreading front areas (North Qilian, North Altun and North Kunlun),and thick piedmont molasse is accumulated in the South Tarim-Zoulang sag belt. Spreading of the backland part started in the Triassic. The average spreading rate since the Triassic, estimated on the basis of the amount of strike-slip of the Altum sinistral fault, is 3.2-3.6mm/a for the East Kunlun-Qaidam area and 1.4-1.6mm/a for the Qilian Mountains area. The rate of relative motion between the Alxa massif and Qilian Mountains since the Quaternary, esti-mated from the nappe structures in the Zoulang belt, is 1.35-7.50mm/a. The clockwise rotation of the Kunlun-Altun-Qilian massif and the intense sinistral strike-slip of the Altun fault has led to a compression-extension transition in the Hoh Xil area since the Triassic and the intracontinental tension and rift-type magmatism since the Miocene[6].

Qinghai-Tibet Hinterland: Deep Thermal-uplift – Extension and E-W Asymmetric Spreading.
The uplift mechanism and dynamic process in the hinterland of the Qinghai-Tibet plateau are manifested mainly by vertical extension and E-W asymmetric spreading. Major deformation styles of the vertical extension are: thermal-uplift – extension, structural diapirism and fault blocking. In the hinterland there are four thermal-uplift – extension zones (or layers) at different structural levels: the mantle diapir zone (or mantle plume)at the depth of 200-300km in the Hoh Xil-Bayan Har area; the asthenospheric upwelling zone at the depth of 100-130km in the Qinghai-Tibet Hin-terland; the lithospheric mantle diapir zone at the depth of 55-60km in the Gangdise-Qiangtang area;and the N-S intracrustal detachment layer buried at the average depth of 15-20km and travers-ing the entire orogenic complex (Fig.3). The effect of the thermal-uplift – stretching on the litho-spheric displacement field is manifested on the whole by upward and lateral migration of mate-rial,of which the amount increases with the geothermal gradient [6]. The intense N-S contraction at the west margin (KaraKorum and Pamirs)of the orogenic complex has resulted in the eastward lateral flow or spreading of material in the hinterland(Fi g.4).In the Gangdise-Qiangtang area south of the Xijir Ulan-Jinsha fault, however, the east-spreading regional displacement field is superim-posed by a local displacement field spreading symmetrically east and west.In the Hoh Xil-Bayan Har area the east-spreading regional displacement field is overprinted by a south-spreading local displacement field (Fig.4). Regional spreading in the hinterland of the Qinghai-Tibet plateau com-menced in the terminal Late Cretaceous-EarlyCenozoic. Since the Miocene Pliocene the hinterland was in a stage of strong spreading. Local spreading mainly occurred since the Latest Pliocene (3Ma). Materials in the hinterland are mainly accomodated within the Tethyan-Himalayan orogenic complex, except that a part of them are overthrust on theYangtze massif through the piedmont progressive thrusting at the Longmen Mountains[18] and another part are consumed outside the plateau by the ENE dextral strike-slip faults at the Longmen Mountains and a series of NNW to N-S dextral trike-slip faults in the Sanjiang (Three Rivers)-Western Yunnan area[Fig.4]. It is prelimi-narily estimated that the E-W average spreading rate in the Gangdise-Qiangtang area is 5-19mm/a, higher than that (4.9mm/a) in the Himalayan foreland belt, and the eastward average spreading rate in the Hoh Xil-Bayan Har area is 12.5mm/a(Fig.4).

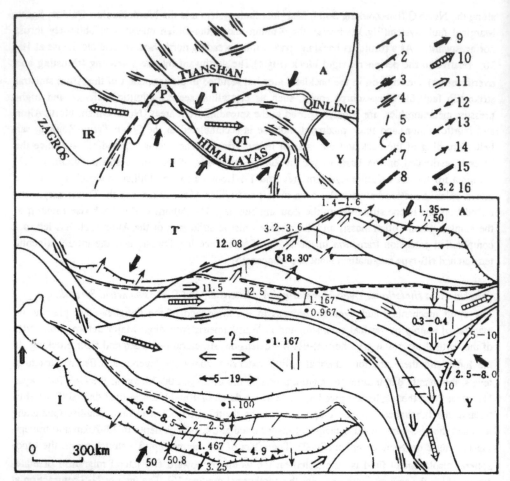

Figure 4. Dynamic sketch showing Tethyan-Himalayan orogenic complex and its nearby areas. 1,thrust; 2,normal fault; 3,sinistral strike-slip fault; 4,dextral strike-slip fault; 5,stretching lineation; 6,rotating direction and angle of massifs; 7,spreading front belt; 8,spreading root belt and spreading boundary of strike-slip type; 9,underthrusting direction; 10,regional movement (spreading)direction;11,local movement (spreading)direction; 12,converging direction between Indian plate and Eurasia plate; 13,stretching (or spreading)direction in massifs; 14,overthrusting direction; 15,spreading boundary of thrust type; 16,average uplifting rate of massifs(since 3 Ma);unit:mm/a.upper figure showing relative motion between Himalayan orogenic complex and its nearby massifs. IR, Iran plateau; QT,Qinghai-Tibet plateau; P,Pamirs; I,India massif;Y,Yangtze massif; T,Tarim massif; A, Alxa-Ordos massif.

CONCLUSINOS

Types of Lithospheric Fabric

Based on the above-mentioned analysis, the characteristics of the three types of lithospheric fabric can be summarized as follows:

Himalayan thrust – superposition type This type of lithospheric fabric is characterized by a thermal texture of hot crust and cool mantle, a strain field with both contraction and stretching, a stress field with the nearly N-S horizontal σ_1 and the early E-W horizontal σ_3, and a lithospheric structural pattern featured by deep level subduction (contraction) and shallow-level thrusting (spreading).

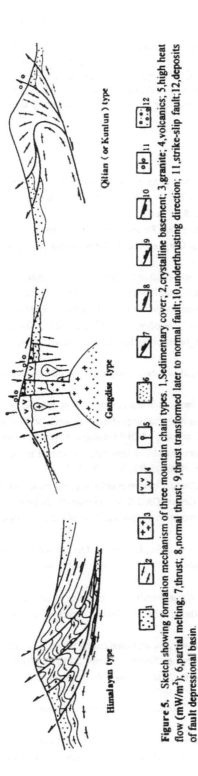

Figure 5. Sketch showing formation mechanism of three mountain chain types. 1,Sedimentary cover; 2,crystalline basement; 3,granite; 4,volcanics; 5,high heat flow (mW/m²); 6,partial melting; 7,thrust; 8,normal thrust; 9,thrust transformed later to normal fault;10,underthrusting direction; 11,strike-slip fault;12,deposits of fault depressional basin.

Qinghai-Tibet strike-slip – stretching type

This type of lithospheric fabric is further divided into two subtypes: the Northern Tibet strike-slip – thermal-uplift extensional subtype; and the Hoh Xil-Bayan Har strike-slip – thermal-uplift spreading subtype. The former is characterized by a thermal texture of hot crust and hot mantle; a strain field with strike-slip, extension and E-W spreading; a stress field with the nearly vertical σ_1 and the nearly E-W horizontal σ_3; and a lithospheric structural pattern featured by upwelling of the lithospheric mantle and asthenosphere as well as multi-level spreading. The latter is characterized by a thermal texture of cool crust and hot mantle; a strain field with strike-slip, extensional and eastward spreading; a stress field with the nearly vertical σ_1 and the NE horizontal σ_3; and a lithospheric structural pattern featured by upper mantle diaprism and mult-level eastward spreading which in turn is superim posed by southward spreading.

Kunlun-Altun-Qilian thrust – spreading type

This type of lithospheric fabric is characterized by a transitional thermal texture; a strain field with a combination of thrusting and strike-slip, extension and spreading, translation and rotation; a stress field with the NNE horizontal σ_1 and the nearly vertical σ_3 and a lithospheric structural pattern featured by deep-level subduction (contraction) and shallow-level stretching (spreading and extension).

Genetic Types of Mountain Chains

The mountain chains constituting the Tethyan-Himalayan orogenic complex, although having experienced different orogenic cycles, rose mainly in the terminal Pliocene-Quaternary (since 3Ma), suggesting that they are related to each other in uplift- mechanism and have undergone a unified continental dynamic process.On the basis of the uplift mechanism and dynamic process the orogenic complex can be divided into three mountain chain genetic types: the Himalayan,the Gangdise and the Qilian (or Kunlun) (Fig.5).

The uplift mechanism of *the Himalayan type* is dominated by the superimposition of rock slices caused by contraction (thrusting,

152

subduction and folding) and multi-level decoupling and is characterized by an abrupt and unbalanced uplift.That of *the Gangdise type* is represented mainly by thermal-uplift extension and the diapirism of deep-seated rocks and fault blocking and is featured by a rapid and roughly balanced uplift. The Kunlun Mountains, Altun Mountains and Qilian Mountains have a similar uplift mechanism which is called the *Qilian type (or Kunlun type)* for short. Uplift mechanism of this type is marked principally by the deep-level contraction (subduction) and shallow-level extensinal (diapirism and fault blocking) and spreading (thrusting and strike slipping) and is featured by a slow and approximately isostatic uplift.

REFERENCES

1. Cui Junwen,Wu Changde,Zhu Hong,Deng Zongce and Feng Xiaofeng. Mechanism for continental-crust thickening and uplifting of the Himalayan collision belt — A model for continental-crust tectonic evolution.Bulletin of the Chinese Academy of Geological Sciences, 21,55-64(1990).
2. Cui Junwen,Zhu Hong,Wu Changde,Feng Xiaofeng,Tan Zheming and Deng Zongce.Yadong Golmud GGT,deformation and dynamic of the lithosphere in Qinghai-Xizang (Tibet)plateau.Geological Publishing House,Beijing(1992).
3. Cui Junwen.Deep-level expansion of the lithosphere and uplift of the Qinghai-Tibet plateau — A discussion of the lithosphere deep-level expansion models. Geological Review.   40(2),106 - 110(1994).
4. Cui Junwen,Wu Changde and Zhu Hong.The tectonic evolution of the Qinghai- Tibet plateau and the uplift mechanism of the Himalayas. In: Shi Yangshen, Lu Huafu, Ma   Ruishi and Sun Yan(eds.).Symposium of the researches on modern geology(Vol.II), Nanjing University Publishing House, Nanjing,127 - 136(1994).
5. Cui Junwen and Xu Zhiqin. The Tethyan-Himalayan orogenic complex.   Institute of Geology, CAGS, Continental Dynamics, 1(1),20 - 29(1996).
6. Cui Junwen, Shi Jinsong, Tang Zheming, Dai   Junming.   Wang Lianjie.   Post-orogenic deformation of the Qinghai-Tibet plateau and its uplifting mechanisms. Metallurgical Industry press, Beijing(1997).
7. Cui Junwen.Tectonic evolusion of Himalayan collision belt.   ACTA Geologica Sinica(in press).
8. Cui Zuozhou, Li Qiusheng, Wu Chaodong,Yin Zhorxun and Liu Hongbing.The crustal and deep structure in Golmud-Ejin Qi GGT.ACTA Geophysica Sinica .38(suppl. II),15 - 28 (1995).
9. Dong Xuebin ,Wang Zhongmin, Yang Huixin and Cheng Liren. New paleomagnetic results and amalgamative relations of terranes.In: Lithosphere Research Centre, CAG S and Institute of Geology, CAGS (Eds.), Yadong Golmud GGT,Lithospheric structure and evolution of Qinghai-Tibet plateau. Geological Publishing House, Beijing, 7 6 - 82(1996).
10. A. Hirn, A Nercessian, M Sapin, G Jobert, Xu Zhongxin,Gao Enyuan, Lu Deyuan and Teng Jiwen. Lhasa block and bordering sutures — A continuation of a 500 km Moho traverse through Tibet.Nature,307(3946),25 - 27(1984).
11.Meng Lingshun,Gao Rui,Zhou Fuxiang, Li Li and Wang Huaxiao. Interpretation of the crustal structure in Yadong-Golmud area using gravity anomalies. Bulletin of the Chinese Academy of Geological Sciences,21,149 - 161(1990).
12. Meng Lingshun, Guan Ye, Qi Li and Gao Rui. Gravity field and deep crustal structure in Golmud-Ejin Qi geoscience transection and nearby area.ACTA Geophysica Sinica, 38(Suppl. II),36 - 45(1995).
13. Shen Xianjie,Yang Shuzhen, Shen Jiying.Heat flow study and analysis along the Golmud-Ejin Qi geotransect.ACTA Geophysica Sinica, 38(Suppl. II),86 - 97(1995).
14. P.Tapponnier, J.L.Mercier, P.F.Proust et al.. The   Tibetan side of the India   Eurasia collision. Nature, 294, 405 - 410(1981).
15. Wang Wenying. The crustal movement and its geodesy of the Qinghai-Xizang (Tibet) plateau. In: China Society of the Qinghai-Tibetan Plateau Research (eds.) ,Qinghai-Tibetan plateau and global variations,

China Meteorological press, Beijing, 214 – 221 (1995).

16.Wu Gongjian,Xiao Xuchang, Li Tingdong, Cui Junwen and Gao Rui. Global geoscience transect 3, Yadong to Golmud transect, Qinghai-Tibet plateau, China. American Geophysical Union(1991).

17.Wu Xuanzhi, Wu Chunling, Lu Jie, Wu Jie.Research on the fine crustal structure of the Northern Qilian-Nexi Corridor by deep seismic reflection.ACTA Geophysica Sinica. 38(Suppl. II),29 – 35(1995).

18.Xu Zhiqin, Hou Liwei, Wang Zongxiu, Fu Xiaofang and Huang Minghua. Orogenic processes of the Song-pan-Garze orogenic belt of China.Geological Publishing House,Beijing(1992).

19.Xu Zhiqin,Cui Junwen and Zhang Jianxin.Deformation dynamics of continental mountain changes. Metal-lurgical Industry Press, Beijing (1996).

20.Xu Zhiqin, Jiang Mei and Yang Jingsui. Tectonophysical process at depth for the uplift of the northern part of the Qinghai-Tibet plateau.ACTA Geologica Sinica,70(3),195 – 206(1996).

21.Zhang Shengye,Wei sheng,Wang Jiaying. Investigation of electrical structure of the crust and upper mantle in the Qiangtang basin of Tibet. Annual of the Chinese Geophysical Society,85(1995).

22.Zhu Renxue,Hu Xiangyun.Study on the resistivity structure of the lithosphere along the Golmud-Ejin Qi geoscience transect.ACTA Geophysica Sinica. 38(Suppl. II), 46 – 57(1995).

Proc. 30ᵗʰ Int'l. Geol. Congr., Vol. 6, pp. 155-168
Xiao Xuchang and Liu Hefu (Eds)
© VSP 1997

Petrology and Geochemistry variations of Mesozoic and Cenozoic volcanism of the Tibetan Plateau and Its Dynamical Inferences for Lithospheric Evolution of the Plateau

SHUANGQUAN ZHANG[1], XUANXUE MO, CHONGHE ZHAO, TIEYING GUO and WAN JIANG

Department of Geology, China University of Geosciences, Beijing 100083, P. R. China

Abstract

It is believed that Mesozoic and Cenozoic volcanic rocks of the Tibetan Plateau has well recorded the dynamical processes involving Tethysan subduction, the collision of Indian and Eurasian plates, and subsequently crustal thickening and uplifting of the plateau. Therefore, detailed studies regarding the time-space distribution, petrology and geochemistry variations on Mesozoic and Cenozoic volcanic rocks of the plateau, have been conducted. Furthermore, some inferences for lithospheric evolution of the plateau, are generally discussed.

Keywords: volcanic rocks, Tibetan Plateau, Mesozoic, Cenozoic, pre-collision, post-collision, uplifting, crustal thickening

INTRODUCTION

At present, a lot of detailed data [9-12, 16, 35, 22, 8, 26, 20-21, 34, 23, 2, 29, 19, 36] have been accumulated on Mesozoic and Cenozoic volcanic rocks of the Tibetan Plateau. Based on the basic petrological assemblages of the volcanic rocks and their geochemical features, some authors mentioned above, have to some extent discriminated their forming tectonic settings. Nevertheless, we are still faced to a big challenge how well to understand the deep processes of the crust-mantle system, controlling the surficial uplifting and the forming of double crustal thickness of the plateau. Methodologically, it is believed that volcanic rocks resulted from melting of the lower crust or the mantle rocks, are good indicators of the changing inner earth, and herein, compared with geophysical techniques, they are more effective to reveal the evolution of the crust-mantle system beneath the Tibetan Plateau. Furthermore the researches on the Mesozoic and Cenozoic volcanic rocks, have proved to be able to make great clues to the interpretations of the uplifting and the crustal thickening of the plateau, so called "the biggest tectonic puzzle of Cenozoic earth"[28].

With this regard, based on the previous works mentioned above, and our new data on the Mesozoic and Cenozoic volcanic rocks from the Lhasa area [36], in this paper we not only

---

[1] Now at the Department of Geology, Peking University, Beijing 100871, P. R. China

156

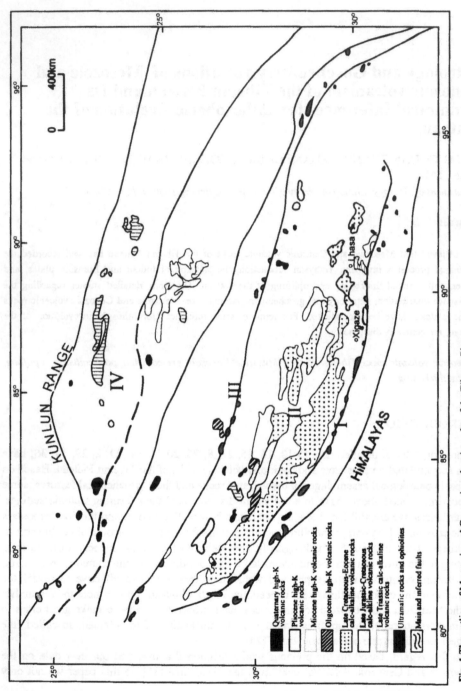

Fig. 1 The outline of Mesozoic and Cenozoic volcanic rocks of the Tibetan Plateau.
I=Yalung Zangbo oceanic basalt and ophiolite belt (late Jurassic-early Cretaceous), II=Gangdise calc-alkaline volcanic Province (Late Jurassic-Neogene),
III=Bangongco Nujiang basalt and ophiolite belt (Jurassic), IV=North Tibet volcanic Province (Miocene-present-day).

wholly summarize time-space distribution and tectonomagmatic types, regional variations of petrology and geochemistry of the Mesozoic and Cenozoic volcanic rocks of the Tibetan Plateau, but also discuss their dynamical inferences for the lithospheric evolution of the plateau.

SPACE-TIME DISTRIBUTION AND TECTONOMAGMATIC TYPES OF MESOZOIC AND CENOZOIC VOLCANIC ROCKS OF THE TIBETAN PLATEAU

Although volcanism of the Tibetan Plateau has taken place during a quite long period of geological time, approximately from Carboniferous to present-day, the Mesozoic and Cenozoic volcanic rocks have been thought of as important probes to reveal the dynamical processes of lithospheric evolution of the plateau, involving the subduction of Tethysan oceanic crust, the subsequent colliding between Indian and Eurasian plates and the resultant uplifting and crustal thickening of the plateau. According to Regional Geology of Xizang (Tibet) Autonomous Region [32] and related published data [11, 21], Fig.1 shows the outline of space-time distribution and tectonomagmatic types of Mesozoic and Cenozoic volcanic rocks of the Tibetan Plateau, from south to north, as follows: (1) Yarlung Zangbo belt (I), an important part of Tethysan suture, consists of middle oceanic ridge basalts (MORB) and ophiolites, produced during the period of late Jurassic-early Cretaceous, (2) Gangdise volcanic province (II), covering the biggest area of the volcanic rocks of the plateau, (approximately 77041km^2), is constituted of subduction-related calc-alkaline and small volume of high-K volcanic rocks, which are produced by the Tethyan northward subduction (late Jurassic-Eocene) and the subsequent collision of the two continental plates, as well as the collapse of orogen (since Oligocene), (3) Bangongco Nujiang belt (III) is recognized as another Tethys-related MORB and ophiolites zone, representing one of the late Jurassic oceanic basins or back-arc basins [1], and (4) northern Tibet volcanic province (IV), is dominated by Cenozoic volcanic rocks, mostly characterized by high contents of potassium and total alkalies. Additionally, it is noted that the Cenozoic (specially since Miocene) volcanic rocks, mainly occurring in Northern Tibet and the Gangdise area, are presumably resulted from the processes in connection with the crustal thickening and uplifting of the plateau.

In addition, combined with the intrusive rocks from Tibet, the basic characteristics on space-time pattern of post-Cretaceous magmatism in the plateau, are described as follows: (1) in the Himalayan area, the magmatism is characterized by a suit of leucogranitic rocks aged around 20 Ma; (2) in the Gangdise area, intrusive and extrusive rocks are well coupled in space and time; and (3) in northern Tibet granitoids are rarely found, while high-K Cenozoic volcanic rocks intensively occur , commonly exposed in small scale.

PETROLOGY AND GEOCHEMISTRY VARIATIONS OF THE VOLCANISM OF THE TIBETAN PLATEAU

In a general viewpoint, the collision of Indian and Eurasian took place around 45-50 Ma, hereby Mesozoic and Cenozoic volcanic rocks of the Tibet Plateau are divided into pre-collision and post-collision suit, respectively. Thus, the petrological and geochemical evolutionary trends of the volcanic rocks are exhibited by a series of variation diagrams regarding maijor elements, trace elements, rare earth elements and isotopes.

Major elements

Three suits of major element diagrams (Figs. 2, 3, 4) show big differences between the pre- and post-collision volcanism. In Na_2O+K_2O vs. SiO_2 diagrams showing the chemical evolutionary series of magmatic suit (Fig. 2), the volcanic rocks of the plateau obviously differ from each other. The pre-collision volcanic rocks are predominantly subalkaline series (Fig. 2a) and the previous results [35, 8, 36] have identified them as calc-alkaline series. Post-collision volcanism is different between Gangdise and northern Tibet: (1) Among Gangdise post-collision volcanic rocks, approximate 50% belong to subalkaline series and the rest alkaline series (Fig. 2b), (2) for the Neogene volcanism in northern Tibet, over 70% samples fall into subalkaline series and about 30% alkaline series (Fig. 2c), and (3) the Quaternary volcanic samples dominantly belong to alkaline series (Fig. 2d).

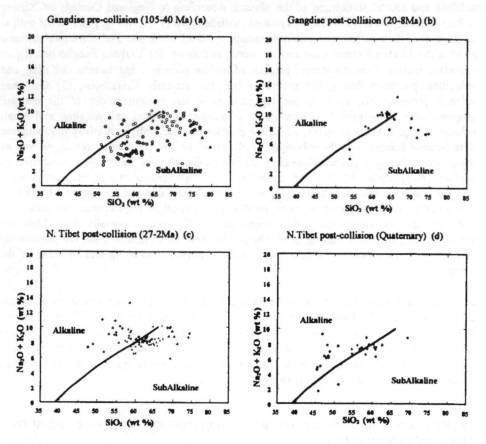

Fig. 2. Na_2O+K_2O vs. SiO_2 diagrams (after Irvine & Barager [15]) for Mesozoic and Cenozoic volcanic rocks of the Tibetan Plateau. The volcanic rocks are grouped into Pre-collision (a) and post-collision (b, c, d). Data for pre-collision group are from [8, 21, 26, 35], and filled triangles represent the samples from the Lhasa area collected in 1994. Data sources for post-collision group are: Gangdise [8, 19], Northern Tibet [2, 9-12, 29].

In the Na_2O+K_2O vs. SiO_2 classification diagrams (Fig. 3), the samples of pre-collision volcanism fall into quite a wide compositional range (Fig. 3a), varying from basic (basalt) to acid (rhyolite) along the SiO_2 axis, from silica-saturated to undersaturated types along the Na_2O+K_2O axis; on the other hand, the post-collision ones exhibit a relatively small composition change: (1) in the Gangdise area, they mainly plot in the fields of trachyte and rhyolite (Fig. 3b), and (2) in northern Tibet, Neogene volcanic rocks mostly display as trachyandesite and trachyte (Fig. 3c), while Quaternary volcanism as trachybasalt and trachyandesite (Fig.3d). In addition, there are some silica-undersaturated types in the plateau.

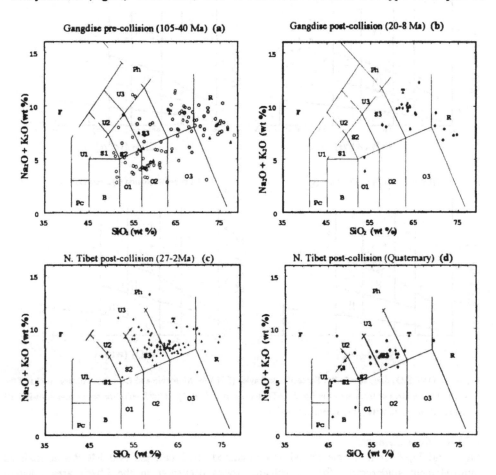

Fig. 3. Na_2O+K_2O vs. SiO_2 classification diagrams (after Le Bas et al. [18]) for Mesozoic and Cenozoic volcanic rocks of the Tibetan Plateau. Data sources and the sample symbols are as in Fig. 2. Symbols: F, foidite; Pc, Picrobasalt; B, basalt; O1, basaltic andesite; O2, andesite; O3, dacite; S1, trachybasalt; S2, basaltic trachyandesite; S3, trachyandesite; T, trachyte (trachydacite); R, rhyolite; U1, tephrite (basanite); U2, phonotephrite; U3, tephriphonolite; Ph, phonolite.

K_2O vs. SiO_2 diagrams may provide important information on the genesis and evolution of volcanic rocks. K_2O vs. SiO_2 diagrams (Fig. 4) show giant differences between pre- and post-collision volcanism. Some conclusions are drawn as follows: (1) Gangdise pre-collision volcanic rocks range between low-K calc-alkaline series and shoshonitic series (Fig. 4a), (2) limited samples of Gangdise post-collision volcanic rocks mostly fall into the fields of high-K calc-alkaline and shoshonitic series (Fig. 4b), (3) Neogene post-collision volcanic samples from northern Tibet mainly plot as shoshonitic series (Fig. 4c), and (4) the Quaternary volcanism predominantly exhibits as shoshonitic sreies (Fig. 4d).

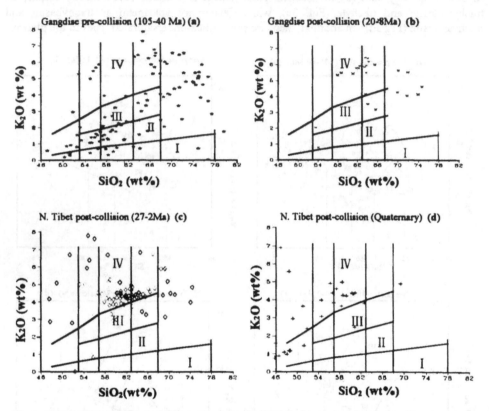

Fig. 4. K₂O vs. SiO₂ diagrams (after Peccerillo & Taylor [27]) for Mesozoic and Cenozoic volcanic rocks of the Tibetan Plateau. Data sources are as in fig. 2. I=Tholeiite series (low-K), II =calc-alkaline series (medium-K), III =high-K calc-alkaline series, IV =shoshonitic series.

REE and trace elements
Fig.5 and Fig.6 demonstrate chondrite-normalized REE patterns and MORB-normalized multi-element spidergrams for the volcanic rocks occurring in the Lhasa area (eastern Gangdise) and northern Tibet, respectively. Light REE (LREE) and large-ion lithophile elements (LILE) abundances in the samples of pre-collision andesites are generally lower than those in the samples of post-collision volcanic rocks from northern Tibet. Although the REE patterns of both pre- and post-collision volcanic rocks are consistently LREE enriched, the former have obvious negative Eu anomalies and relatively small (moderate) slopes (Figs. 5a, 6a). Moreover, apparent Nb-Ta and Ti negative valleys of pre-collision volcanic rocks, are firmly related to oceanic crust subduction; whereas the same cases of Cenozoic volcanic rocks from northern Tibet, presumably infer the subduction of an ancient massif, as coincides with Xie et al's [31] isotope result.

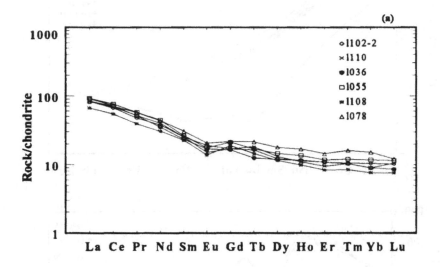

(a)

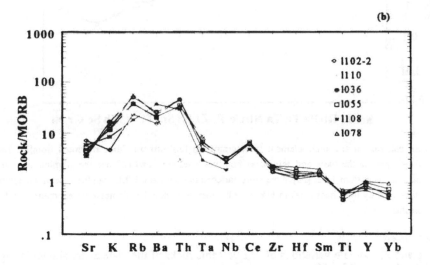

(b)

Fig. 5. REE patterns (a) and multi-element spidergrams (b) for late Cretaceous-Eocene (105-40 Ma) andesites in the Lhasa area, central Gangdise. Note medium enrichment in LREE and LILE, small Eu negative anomalies, Nb-Ta and Ti negtive anomalies, which show typical features of subduction-related volcanic rocks.

162

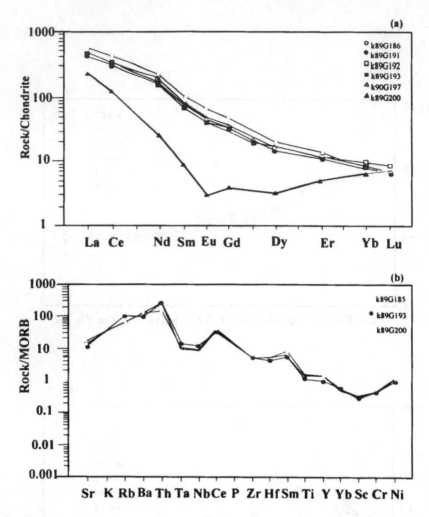

Fig. 6. REE patterns (a) and multi-element spidergrams (b) for northern Tibet, reprinted from[2]. The continuous lines represent the basic and acid lavas from northwestern Tibet [2], and the stippled areas are normalized patterns for eastern Tibet [23]. Note more enriched in LREE and LILE than the case of Gangdise andesites (represented by the Lhasa area) in Fig. 5. Also they just show Nb-Ta negative anomalies, no Eu negative anomalies.

Sr Nd Pb isotopes

The Sr, Nd and Pb isotopic variations of the volcanic rocks of the plateau are shown in Fig. 7. In $^{143}Nd/^{144}Nd$ vs. $^{87}Sr/^{86}Sr$ diagram (Fig. 7a), five samples of Mesozoic and Cenozoic volcanic rocks from the Lhasa area, composed of andesites and rhyolites plot near the mantle array, inferring that mantle material has mainly contributed to the magmas; furthermore, using Langmuir [16] Sr-Nd two components mixing model, we have evaluated 10-20 wt% isotopic additions of continental crust to the magmas. On the other hand, Cenozoic volcanic rocks from northern Tibet are more enriched, far from the mantle array. Additionally, in the $^{208}Pb/^{204}Pb$ vs. $^{206}Pb/^{204}Pb$ and $^{207}Pb/^{204}Pb$ vs. $^{206}Pb/^{204}Pb$ diagrams (Figs. 7b, 7c), the Lhasa volcanic rocks display wider variations along both vertical and horizontal axis than those of northern Tibet which stay quite constant in $^{206}Pb/^{204}Pb$ ratios. Consequently, it follows that pre- and post-collision volcanism is respectively resulted from different source rocks and their evolutionary processes are quite distinguishable.

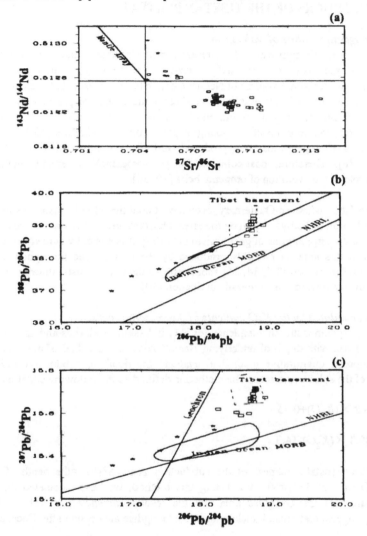

Fig. 7. Nd-Sr and Pb variation diagrams for Mesozoic and Cenozoic volcanic rocks of the Tibetan Plateau. data sources are: Lhasa area [36], Northern Tibet[2, 12, 29, 31]. ★ represents late Cretaceous-Eocene volcanic rocks in the Lhasa area; □ represents Cenozoic volcanic rocks in N. Tibet. Dashed and and continuous lines outline the Tibet basement and Indian Ocean MORB, respectively. Obviously, the volcanic rocks from the Lhasa area markedly differ from those of the northern Tibet in isotopic composition; the former are more depleted than the latter.

DISCUSSION OF DYNAMICAL INFERENCES OF MESOZOIC AND CENOZOIC VOLCANIC ROCKS OF THE TIBETAN PLATEAU

Tectonomagmatic outline of volcanism
In summary, in the Gangdise area, Late Jurassic-early Cretaceous and late Cretaceous-Eocene calc-alkaline volcanism is undoubtedly linked to the Yarlung Zangbo Tethys northward subduction; whereas small volume of Neogene calc-alkaline intermediate volcanic rocks (e. g. Maquiang area), high-K calc-alkaline and shoshonitic rocks are presumably related to the collapse of Gangdise orogen, as currently becomes more critical issue. Notably, in the Maquiang area, the eruption of the youngest potassium calc-alkaline (15-10 Ma) lavas are characteristics of orogenic magmas, produced well after the collision event and cessation of subduction [8]. Therefore, post-collision (or post-orogenic) volcanism should plays an important role in the evolution of orogenic belts [29, 30].

In northern Tibet, Miocene-Quaternary, even recent continental volcanism was rarely effected by the subduction of Tethys. Based on the latest observations for the volcanic rocks [12, 29, 2] and some new interpretations of geophysical data [24], it is basically agreed that the Cenozoic volcanism in this area, is probably produced by the catastrophic lithosphere thinning, i.e. lithosphere delamination [3-6, 14, 25] subsequent to the major crustal thickening, and partly related to the southward Tarim massif subduction [10].

Thickness calculated by the K_2O contents of the volcanic rocks
Dickinson [13] demonstrated a tendency for K_2O in basaltic andesites and andesites at 55-60 wt% to increase with depth of underlying Benioff zone in individual island arcs. Condie [7] has developed a quantitative method to evaluate the depth of subduction and the crustal thickness of the subduction zone. The method is defined as the following equations:

$$C(km)=18.2 \times (K_2O)+0.45$$

$$SZ(km)=89.3 \times (K_2O)-14.3,$$

where $C(km)$=crustal thickness of the subduction zone, $SZ(km)$=the depth of subduction, $K_2O=K_2O$ wt% at SiO_2=60 wt%. Using this method, we have calculated the $C(km)$ and $SZ(km)$ values of the Gangdise area and the results are shown in Tab. 1. Our results essentially suggest that crustal thickening in the Gangdise area began after Eocene.

Tab.1 The variations of crustal thickness and the depth of subduction in the Gangdise area (Triassic-Miocene)

Geo. Time	Sample Number	Area	SiO_2 (wt%)	K_2O (wt%)	Average crustal thickness (km)	Average subduction depth (km)
T -K	22	Lhasa-Anduo	54-62	0.85-3.24	36	164
K_2-E_2	30	Lhasa	53-63	0.79-4.98	37	167
E_3-N_1	3	Gangdise	53-59	0.79-5.48	52	239

Note T=Triassic, K=Cretaceous, K_2=Late Cretaceous, E_2= Eocene, E_3=Oligocene, N_1=Miocene. Samples for the calculation are from[21, 26, 35, 36].

General outline for the lithospheric evolution based on the volcanism of the Plateau
Detailed volcanism studies have suggested a general outline for lithospheric evolution of the Plateau: (1) late Triassic-early Eocene is the period of multi-cycle subduction activities, as is proved by previous study [26]; (2) during late Eocene-middle Miocene, orogenic collapse and resultant extentional activities are dominant in the Gangdise area; and (3) within middle Miocene-Quaternary, lithospheric thinning or delamination subsequent to the crustal thickening, played a main role throughout the most parts of the plateau.

The impacts of magmatism to crustal thickening and uplifting of the plateau
Wilson [33] has pointed out that subduction-related magmatism is one of the most important ways of continental growth, in terms of the lateral accretion of island-arc and intermediate extrusive and intrusion in the vertical direction. The impacts of Mesozoic and Cenozoic magmatism to crustal thickening and uplifting in the plateau, especially in the Gangdise area, lie in two aspects:

(1) Magmatism has directly transported a huge amount of mantle material to the crust, and increased the crustal thickness. Approximately evaluating shows that the andesites in the Gangdise area added about 2.5km-3km thickness to the crust.

(2) Magmatism has brought a great deal of heat to the crust, which is probably a main energy source of the uplifting of the plateau. With regard to this viewpoint, a general energy model for the plateau system can be delineated as the equation,

$$\frac{1}{2} m v^2 + Q = E_h ,$$

where $\frac{1}{2} m v^2$ stands for the horizontal and vertical movements; Q represents the heat mainly caused by magmatism; E_h represents the potential resulted from the change of the elevation of the plateau.

CONCLUSIONS

From south to north, the outline for the Mesozoic and Cenozoic volcanic rocks of the Tibetan Plateau consists of Yarlung Zangbo MORB and ophiolite belt (late Jurassic-early Cretaceous), Gangdise subduction-related calc-alkaline and post-collision volcanic province (late Jurassic-Neogene), Bangongco Nujinag basalt and ophiolite belt (Jurassic), and northern Tibet high-K and shoshonite volcanic province (Cenozoic).

Petrological and geochemical characteristics exhibit significant differences between Pre- and post-collision volcanism. The former is characterized by higher potassium contents, more enriched in LREE and LILE and higher initial Sr and Pb isotopic ratios than the latter. Thus, it follows that pre- and post-collision volcanism represents different dynamical mechanism, respectively; and it is suggested that the former corresponded to the northward subduction of

Tethys, and the latter was essentially controlled by the post-collision extentional activities, including post-orogenic collapse, and lithospheric thinning or delamination.

Crustal thickness evaluated through the K_2O contents of Mesozoic and Paleozoic volcanic rocks from the Gangdise area, suggest that crustal thickness was markedly increased after the collision of Indian and Eurasian, as disagrees with the point that the crust was thickened prior to the collision.

Mesozoic and Cenozoic volcanic rocks of the Tibetan Plateau directly made contributions to the uplifting and crustal thickening of the plateau, including both mass addition and energy delivering. Also, we reasonably proposed an equation (mentioned above) to describe the exchange and transport of the energy in the plateau system, so that the mechanism concerning the uplifting and crustal thickening of the Tibet, can be better understood.

Finally, our works imply that in the Gangdise area, in spite of small volume of high potassium and shoshonitic intermediate rocks mainly dated in 20-8 Ma [8, 19], occurring in the basins and grabens, should provide substantial evidence to the study on the post-collision or post-orogen tectonic mechanism of Tibet.

Acknowledgments

This work was supported by the Ministry of Geology and Mineral Resource program on the Mesozoic and Cenozoic magmatism and tectonic evolution of Tibet. Authors thank Professor Xiao Xuchang and Professor B. C. Burchfiel for their encouragement to present our results in the 30th International Geological Congress, and invitation to write this manuscript for the 30th IGC proceedings. Thanks also are due to Professor Ba Dengzhu and others in Xizang (Tibet) Bureau of Geology and Mineral Resource, for their supports to our field work.

REFERENCES

1. C.J. Allègre, V. Courtillot and 33 others. Structure and evolution of the Himalayas-Tibet orogenic belt, *Nature* 307, 17-22 (1984).
2. N. O. Arnaud, Ph. Vidal, P. Tapponnier, Ph. Matte and W. M. Deng. the high K_2O volcanism of northern Tibet: geochemistry and tectonic implications, *Earth Plant. Sci. Lett.* 111, 351-367 (1992).
3. P. Bird. Initiation of intracontinental subduction in the Himalaya, *J. Geophys. Res.* 83, 4975-4987 (1978).
4. P. Bird. Continental delamination and colorado Plateau, *J. Geophys. Res.* 84, 4975-4987 (1979).
5. P. Bird. and J. Baumgardner. Steady propagation of delamination events, *J. Geophys. Res.* 86, 4891-4903 (1981).
6. P. Bird. Lateral extrusion of lower crust from under high topography, in the isostatic limit, *J. Geophys. Res.* 96, 10,275-10286 (1991).
7. K.C. Condie. Plate tectonics & crustal evolution, Pergamon Press, New York (1976).
8. C. Coulon, H. Maluski, C. Bollinger and S. Wang. Mesozoic and Cenozoic volcanic rocks from central and southern Tibet: $^{40}Ar/^{39}Ar$ dating, Petrological characteristics and geodynamical significance, *Earth Planet. Sci. Lett.* 79, 281-302 (1986).

167

9. W.M. Deng. A preliminary study on the petrology and petrochemistry of the Quaternary volcanic rocks of northern Tibet autonomous region (in Chinese), *Acta Geol. Sinica* 50, 148-162 (1978).

10. W.M. Deng. Cenozoic volcanic rocks in the northern Ngari district of Tibet (Xizang), discussion on the concurrent intracontinetal subduction (in Chinese), *Acta Petro. Sinica* 3, 1-2 (1989).

11. W. M. Deng. Geology, geochemistry and age of shoshonitic lavas in the central Kunlun orogenic belt (in Chinese), *Scientia Geologica Sinica* 3, 201-213 (1991).

12. W.M. Deng. Study on trace element and Sr, Nd isotopic geochemistry of Cenozoic potassic volcanic rocks in north Tibet (in Chinese), *Acta Petro. Sinica* 9, 379-387 (1993).

13. W.R. Dickinson. Circum-Pacific andesite types, *J. Geophys. Res.* 73, 2261-2269 (1968).

14. G.A. Huouseman, D.P. McKenzie and P. Monlar. Convective instability of a thickened boundary layer and its relevance for thermal evolution of continental convergent belts, *J. Geophys. Res.* 86, 6135-6155 (1981).

15. T.N. Irvine and W.R.A. Barager. A guide to the chemical classification of the common volcanic rocks, *Canadian Journal of Earth Sciences* 8, 523-548 (1971).

16. C.W. Jin. Tibetan (Xizang) Volcanic rocks. In: *Magmatism and metamorphism in Tibet area (Academica Sinica Scientific Geotraverse of the Qingzang (Tibet) Plateau result)* (in Chinese). pp. 213-263. Science Publishing House, Beijing (1981).

17. C.H. Langmuir, R.D.Jr. Vocke, G.N. Hanson and S.R. Hart. A general mixing equation with application to Icelandic basalt, *Earth Planet. Sci. Lett.* 37, 380-392 (1978).

18. M.J. Le Bas, R.W. Le Maitre, A. Strecheisen and B. Zanettin. A chemical classification of Volcanic rocks base on the total alkali-silica diagram, *J. Petrol.* 27, 745-750 (1986).

19. J. Li, Y. Zhang and H. Luo. A research on Petrological characteristics and genesis of the Cenozoic volcanic rocks in the Yangying village geothermal field, Dangxiong, Tibet, China (in Chinese), *Geoscience(Journal of Graduate the School of China University of Geosciences)* 6, 96-109 (1992).

20. F. Lu and C. Zhao. Series and characteristics of volcanic suite in Ali, Xizang (Tibet) (in Chinese), *Earth Science (Journal of China University of Geosciences)* 12, 293-300 (1987).

21. F. Lu and C. Zhao. volcanic rocks in the Ngari. in: *Geology of the Ngari Tibet* (in Chinese). T. Guo, J. Liang, C. Zhao and Y. Zhang (eds). pp. 151-198. China University of Geosciences Press, Wuhan (1991).

22. H. Maluski, F. Proust and X.C. Xiao. $^{39}Ar/^{40}Ar$ dating of the Trans-Himalayan calc-alkaline magmatism of southern Tibet, *Nature* 29, 152-154 (1982).

23. L.W. McKenna and J.D. Walker. geochemistry of crustally derived leucocratic rocks from the Ulugh Muztagh area, northern Tibet and their implications for the formation of the Tibetan Plateau, *J. Geophys. Res.* 95 (B13), 21,483-21,502 (1990).

24. P.A. Molnar. A review of geophysical constrains on the deep structure of Tibetan Plateau, the Himalaya, and the Karakorum and thier tectonic implications, *Trans. R. Soc. London ser. A* 326, 33-88 (1988)

25. K.D. Nelson. Are crustal thickness variations in old mountain belts like the Appalachians a consequence of lithospheric delamination ?, *Geology* 20, 498-502 (1992).

26. J.A. Pearce and H. Mei. Volcanic rocks of the 1985 Tibet Geotraverse: Lhasa to Golud, in The Geological evolution of Tibet, *Phil. Trans., R. Soc. London Ser.* 335, 341-345 (1988).

27. A. Peccerillo and S.R. Taylor. Geochemistry of Eocene calc-alkaline volcanic rocks from the Kastamonu area, northern Turkey, *Contrib. Mineral. Petrol.* 58, 63-81 (1976).

28. R. Sorkhabi, A. Macfarlane, G. Mason, J. Quade. " Roof of the earth" offers clues about how our planet was shaped, *EOS* 77, 385-387 (1996).

29. S. Turner, C. Hawkesworth, J. Liu, N. Roggers, S. Kelley and V. Calsteren. Timing of Tibetan uplifting constrained by analysis of volcanic rocks, *Nature* 364, 50-54 (1993).

30. S. Turner, M. Sandiford and J. Foden. Some geodynamic and compositional constraints on " postorogenic " magmatism, *Geology* **20**, 931-934 (1992).

31. G.H. Xie, C.Q. Liu and A. Masuda. The geochemical characteristics of Cenozoic volcanic rocks in the area of Qinghai-Tibet Plateau (in Chinese). In: *The age and geochemistry of Cenozoic volcanic rock in China*. R.X. Liu (ed). pp. 400-427. Seismological Publishing House.

32. Xizang (Tibet) Bureau of Geology and Mineral Resource. *Regional geology of Xizang (Tibet) Autonomous Region* (in Chinese). Geological Publication House, Beijing (1993).

33. M. Wilson. *Igneous petrogenesis, a global tectonic approach*. Uniwin Hyman, London (1989).

34. R.H. Xu, U. Schärer and C.J. Allègre. Magmatism and metamorphism in the Lhasa area block (Tibet), an U-Pb geochronological study, *J. Geology* **93**, 41-57 (1985).

35. S. Wang. Volcanic rocks in central-southern Xizang (in Chinese). In: *Tectonic evolution of the Lithosphere of the Himalayas*. G. Liu et al. (eds). pp. 199-233. Geological Publishing House, Beijing (1990).

36. S.Q. Zhang. *Mesozoic and Cenozoic Volcanic Rocks from Lhasa Area, Central Gangdise: Characteristics and the Significance as Deep Indicators of Lithospheric Evolution in the Tibetan Plateau [unpublished Ph. D. Dissertation]* (in Chinese). China University of Geosciences, Beijing (1996).